MIRACLES AND PHYSICS

By the same author

Les tendances nouvelles de l'ecclésiologie

The Relevance of Physics

Brain, Mind and Computers
(Lecomte du Nouy Prize, 1970)

The Paradox of Olbers' Paradox

The Milky Way: An Elusive Road for Science

*Science and Creation: From Eternal Cycles
to an Oscillating Universe*

*Planets and Planetarians: A History of Theories
of the Origin of Planetary Systems*

The Road of Science and the Ways to God
(Gifford Lectures: University of Edinburgh, 1975 and 1976)

The Origin of Science and the Science of its Origin
(Fremantle Lectures, Oxford, 1977)

*And on This Rock: The Witness of One Land
and Two Covenants*

Cosmos and Creator

Angels, Apes and Men

Uneasy Genius: The Life and Work of Pierre Duhem

Chesterton: A Seer of Science

The Keys of the Kingdom: A Tool's Witness to Truth

Lord Gifford and His Lectures: A Centenary Retrospect

(continued on p. [100])

Miracles
and
Physics

Stanley L. Jaki

Christendom Press

Second, entirely reset edition

© 1999 Stanley L. Jaki

All rights reserved
Manufactured in the United States of America

All inquires should be addressed to:
Christendom Press
134 Christendom Dr.
Front Royal, VA 22630

No part of this book may be reproduced in any form, by any electronic or mechanical means, including information storage and retrieval systems, without permission in writing from the publisher, except by a reviewer who may quote brief passages in a review.

Stanley L. Jaki (1924-)
Miracles and Physics
1. Miracles. 2. Classical and Modern physics

ISBN 0-931888-70-0

Contents

Foreword	vii
Introduction	1
Ch. I. Prior to the Twentieth Century	17
Ch. II. In the Twentieth Century	41
Ch. III. Perennial Perspectives	77
Index of Names	97

The most incredible thing about miracles
is that they happen
*G. K. Chesterton**

The Innocence of Father Brown
(1911; Penguin Books, 1950), p. 11.

Foreword

Since the writing of this essay and its publication as a book in the late 1980s the topic of miracles repeatedly regained my attention. Proofs of this are an essay of mine on Newman's views on miracles,[1] two chapters in my book *Bible and Science*,[2] and a chapter in my latest book *Means to Message: A Treatise on Truth*.[3] The question of miracles is, of course, visible everywhere in my forthcoming book, *God and the Sun at Fatima*.

I thought it best to leave the text of this book essentially unchanged. This way the new edition fully mirrors the state of the author's mind at the time of the writing, although this was a secondary consideration. The primary consideration relates to the perspective of this book, which is the conviction that the basic precepts of epistemology are not a matter of changing fashions. This also holds true of the limitations that are inherent in the scientific method. Dealing as it does ultimately with quantitative properties of material interactions and

---

[1] "Newman and Miracles," *Downside Review* 115 (July 1997), pp. 193-214. It will be reprinted in my *Newman's Challenge* (forthcoming), a collection of my essays on Newman.

[2] Christendom Press, 1996.

[3] Wm. B. Eerdmans, 1999.

nothing more, that method is signally inapplicable to the possibility and purpose of miracles. As to the reality of miracles, physics, this most exact of all sciences, merely provides the backstage against which certain facts reveal the finger of God at play.

Those who tried either to refute or to defend miracles with an eye on classical and modern physics invariably fell prey to the lure of trendy considerations. This book shows at least this much, especially with respect to modern physics. This should seem of no small significance at a time such as ours when too many theologians look for dubious handouts from what they think to be science, although they merely receive thereby a doubtful interpretation of it. They should show more appreciation of the treasure they are supposed to hold in their professional hands and hang on to it for dear life. For the treasure is God's word and miracles are the seal on its revealed character.

December 1998

Introduction

Just about a hundred years ago, Hyppolite Taine, one of the first to pour the interpretation of cultural history into Darwinian moulds, traveled through Italy and did some Church-watching. Then as now Catholicism in Italy offered a unique mixture of decay and saintliness in the midst of a modernity on the rampage. In reflecting on modernity Taine had more in mind than the rapid transformation of life through the onrush of railroads and electrification. As one keen on registering new forms of thinking he wondered aloud whether the Catholic Church would have enough strength to resist theological modernism.

In saying that "if Catholicism resists this attack it seems to me that it will forever be safe from all other attacks,"[1] Taine wanted to compliment modernism rather than a possibly victorious Church. He knew that modernism represented the very essence of what only the Catholic Church opposed as a body and with no sign of readiness to compromise. It was no secret for Taine that theological modernism was the most polished form of a humanism that brooked no interference from any factor

---

[1] Quoted in V. Giraud, *Essai sur Taine, son oeuvre et son influence d'après documents inédits* (2d rev. ed.: Paris: Hachette, 1901), p. 75.

above. He also saw that the Catholic Church represented in a most concrete way the view that such interference occurs all the time and everywhere. Otherwise the Church would not pray and would not enjoin on any and all the sacred duty of daily prayer.

Whatever the possibility of a pantheistic mysticism within modernism, the latter is irreconcilable with prayer insofar as prayer presupposes its being answered by a God different from the world. Hence the opposition of modernism to prayer, properly so called, comes particularly to a head in respect to prayers whereby man tries to obtain strictly miraculous events. Indeed, in reacting to modernism the Church stressed the centrality which miracles play in the Christian dispensation. Through his *motu proprio*, issued on September 1, 1910, against modernism, Pope St. Pius X made it obligatory for Catholics to view miracles as "most certain signs of the divine origin of Christian religion," signs perfectly suited for the understanding "of all ages," including modern times.[2]

Had Taine lived to see the vigor whereby awareness of the supernatural was maintained in the Church throughout the modernist crisis, he might have convinced himself that the Church was forever at safe remove from a new major attack by modernism. But Taine, a secularist empiricist, could not gain a view in depth about things empirical. For it takes a supernatural sensitivity to register the evidence of what is the most empirical of all Christian dogmas, the dogma about original sin. Its ever-present effects guarantee the contin-

---

[2] The pope merely repeated the solemn affirmation at Vatican I concerning the role of miracles in Catholic faith.

ual possibility for a recurrence of the modernist crisis in the Church, especially as she has to thrive in a post-Christian cultural framework.

The essence of post-Christianity is the broad cultural conviction that man has successfully emancipated himself from the clutches of the supernatural. About that self-emancipation only that part needs to be recalled here which relates to the steps whereby post-medieval man has grown deeply suspicious of any and all miracles. It should indeed be most instructive that the first of those steps was taken at the end of the High Middle Ages when awareness of miracles, real and bogus, were a way of daily life. But for some, even the great number of medieval miracles was far from enough. The mystical ferment that came to a boiling point among the "spiritualist" branch of the Franciscans, produced such a craving for miracles as to make any event miraculous. Such is the genesis of Ockham's occasionalism in which God creates anew every moment (occasion) and makes thereby every event a miracle.[3]

Like any surfeit, the surfeit of miracles too had to provoke a reaction that came with the rise of Cartesian rationalism. Not that Descartes had not believed in miracles. He even made a pilgrimage to Loreto, to pray there in the house of the Virgin Mary, which pious belief held to have been miraculously transported from Nazareth. He might have been inspired by an earlier pilgrim, Montaigne, who for all his skepticism avowed: "One finds there more of the real sense of religion than in any

---

[3] For details, see my Gifford Lectures, *The Road of Science and the Ways to God* (Chicago: University of Chicago Press, 1978), pp. 40-43.

other place that I have ever seen."[4] As a philosopher, who wanted to banish not only skepticism but all doubts, Descartes offered about the macrocosmos and microcosmos an explanation within which miracles had at best an uneasy status. No wonder that Descartes' printed works contained no follow-up to his youthful manuscript note[5] in which the Incarnation was singled out as one of God's three great miracles. As to the other two, free will and creation in the beginning, Descartes' dicta were markedly ambiguous. Pascal was very much to the point in noting that Descartes allowed no more to the Creator than an initial fillip that set the world machine into motion.[6]

The boast of many scientists throughout the latter half of the 17th century that the mechanistic world view greatly devolved to God's glory, had a hollow ring. Such a devout Christian among them as Robert Boyle could bring himself to accepting only the biblical miracles. Only there could be no rational defense for such a restriction. With Pierre Bayle the rationalist attack on miracles got in high gear at the start of the 18th century. The Enlightenment had not yet run its course when the attack reached its logical denouement in Hume's skepticism and Kant's apriorism. Hume gained wide acceptance to his contradictory argument against miracles, whereas Kant declared them to be inadmissible by

---

[4] Montaigne, *Journal de voyage en Italie, par la Suisse et l'Allemagne en 1580 et 1581*, ed. Charles Dédéyan (Paris: Société de belles lettres, 1946), p. 260.

[5] Quoted in ch. 2 below.

[6] *Pascal's Pensées*, tr. W. F. Trotter, with an introduction by T. S. Eliot (New York: E. P. Dutton, 1958), #77.

critical reason. Kantian rationalism quickly demolished whatever intellectual confidence Protestant theologians had about miracles.[7] Within Catholicism the work of Benedict XIV on the rules of canonization was a signal example in the midst of the Enlightenment that *facts*, however extraordinary, were not to be doubted on the basis of dubious *ideas*.[8]

When a revulsion for the Enlightenment set in, it was all too often articulated by those who had ambivalent feelings towards the Christian Creed. While they looked for the esthetic solace which the edifices and rites of the Church could provide, the truths of the Creed remained alien to their minds that had been conditioned by French rationalism and German idealism to a greater degree than they were ready to admit. The sympathy of many romantics for pantheism was no accident. Nor was it surprising that as Western civilization became progressively engulfed in material pursuits fueled by technological advances, many nurtured a very special hope. They hoped for a syncretism in which strictly secularist progress and ecclesial solace would go hand in hand

---

[7] Protestantism was overwhelmed by the rationalism and materialism of the Enlightenment, and discussion of the miraculous nearly disappeared," writes Morton Kelsey in his article "Miracles: The Modern Perspectives," *Encyclopedia of Religion* [New York: Macmillan 1987], vol. 8, p. 550), a statement with which it would be difficult to disagree.

[8] Not only the vast and classic work of Benedict XIV but all books by Catholic authors on miracles are ignored by Kelsey according to whom "the Roman Catholic church upheld its practice [of endorsing miracles] without trying to defend it intellectually, and shrines like Lourdes drew great crowds" (ibid). Such a statement betrays either an abysmal ignorance or a disreputable prejudice.

without the need to endorse the Creed with its affirmation of miraculous interventions by God. Modernism within the Church less than a hundred years ago as well as today is the evidence of that hope, if not plain wishful thinking, on the part of many who should know better.

The ravages of modernism in non-Catholic Christian churches are all too well known. Even with full awareness of this, one cannot help feeling a lump in one's throat when matter-of-fact references to Jesus' miracles, such as his multiplication of the loaves, are greeted with condescending smile in scholarly gatherings dominated by main-line Protestant theologians. The clinging of some of them to at least the miracle of Jesus' resurrection has indeed a touch expressive of despair about the rationality of revelation.[9]

Evangelicals, or in general Protestants who want to strike out anew on the road of purely biblical Christianity without repeating some blatant missteps of the Reformers, are a cause of encouragement as well as forebodings. The encouragement is part of a debt which it is my duty to acknowledge. The general liberality of mind with which the editorial board of *The Asbury Theological Journal* invited and accepted this essay on miracles is not something a conservative Catholic thinker can readily expect nowadays from editors of well-established Catholic theological periodicals. The foreboding relates to the outcome of the task with which evangelical Christians still must come to grips. The task is to face up courageously to the kind of philosophy that alone can help one steer clear of the traps of occasional-

---

[9] A most recent example of this is *Christ is Risen* (London: Mowbray, 1988) by Richard Harris, Anglican bishop of Oxford.

ism (and the subjectivism it invites) into which the Reformers had fallen headlong.

Today those traps claim many Catholic victims. It is doubtful that any real premonition was felt about this by J. M. Lloyd Thomas as he reviewed Ronald Knox's *Enthusiasm* in the March 1951 issue of the *Hibbert Journal* and raised a question that may appear prophetic:

> . . . the Roman Church, *tam antiqua et tam nova*, haunted by many recurring conflicts and crises, skilled in statesmanship, enriched by unfathomable memory, is vigilant Mother-Confessor and a wise Director of Souls. A candid Protestant, or an honest humanist will find himself at least asking—what in a world rocking in helpless indecision and revealing ominous cracks of threatened collapse, will become of our christian heritage and traditional culture should that Church compromise its sense of divine commission or if, bribed or tortured by lust of power, it should tremble to impose its own discipline, lose its nerve and snap under breaking strain?[10]

Not that the Church has changed, let alone snapped. Her official documents—including all texts of Vatican II—are a genuine part of a continuity reaching back to the apostles. But one can certainly appreciate the bitter feelings John Henry Newman had on finding that not all priests were paragons of doctrinal firmness.[11] Were he alive today, he would have many bitter pills to swallow. Demythologization of miracles, with blithe references to Bultmann's authority, has become a standard feature of "theology" courses in Catholic seminaries and colleges.

---

[10] *The Hibbert Journal* 49 (1951), pp. 305-06.

[11] Newman was to find that out within six short years following his encomiums of Catholic clergy in his *Apologia*.

At safe remove from any intellectual doubt stands in those places, say, the lengthy and laborious essay of Karl Rahner on the Virgin Birth.[12] There a firm declaration of Mary's virginity *in partu* as part of the Catholic Faith is coupled with the claim that we really do not know what that virginity means. Such is a perfect example of the Kantian dichotomy if not schizophrenia between the noumena and the phenomena, though in reverse. Rahner knows what is unknowable according to Kant, namely, the essence of Mary's virginity *in partu*, though he views it as something not specifiable. At the same time he avoids knowing what can plainly be specified and was indeed specified by many Church Fathers, namely, that the infant Jesus left the body of Mary in the same miraculous way in which the risen Jesus appeared through the closed door.

Theology and exegesis infected with Kantianism has produced a conceptual atmosphere that is impossible not to inhale. Of course, just as it is with the pollution of the air, healthy or even half-healthy lungs will react with spells of violent coughing. An illustration of this is the graffiti that appeared in St. John's University in Brooklyn some time in the Fall of 1988. I learned about it from the big Edinburgh daily, *The Scotsman*, in mid-December. Word about the graffiti got into *The Scotsman* from the *Ministers' Forum*, a Church of Scotland publication, with the remark that it should appeal to "ministers with a sense of humour (quite a few, really)." Rather an appeal

---

[12] "Virginitas in partu," in K. Rahner, *Theological Investigations. Vol. IV. More Recent Writings*, tr. K. Smith (Baltimore: Helicon Press, 1966), pp. 135-62.

INTRODUCTION

to the tragic sense would have been called for by the lines, so poignant in their apparent flippancy:

> Jesus said to them, "Who do you say that I am?" They replied, "You are the eschatological manifestation of the ground of our being, the kerygma of which we find the ultimate meaning in our interpersonal relationships." And Jesus said, "What?"[13]

What the graffiti really suggests is that plain Christians can easily see beneath the veneer of theological sophistication. They instinctively know that Jesus is no longer the true Yehoshua or "Yahweh is Salvation," if He did not match His Father's miracles.[14] Orthodox doctrine about Jesus' divinity or strict equality with His Father has usually included references to Jesus' miracles.[15] Had Vatican II included a more emphatic statement on miracles,[16] there would be now one less reason to apply to it the belated warning of Pope Paul VI that all Councils are one-sided. Indeed, Vatican II has achieved a special status in one-sidedness by the fact that it came to be taken by so many, to recall a recent remark of Cardinal Ratzinger, for an excuse to ignore all previous Councils.[17]

---

[13] *The Scotsman*, Dec. 16, 1988, p. 12.

[14] Christ explicitly draws that parallel following his multiplication of the bread.

[15] As, for instance, in the First and Third Councils of Constantinople. For a handy collection of the most important authoritative declarations of this kind, see R. McInerny, *Miracles: A Catholic View* (Huntington, Ind.: Our Sunday Visitor, 1986), pp. 121-25.

[16] See ibid., p. 123.

[17] Cardinal Ratzinger did so in the Fall of 1988 as he briefed a group of bishops from Chile about the background of the Lefèbre case.

A cause of that development is the fashion of doing theology in terms of "models," as if one could know only the aspects (phenomena) of truth but not the truth (noumenon) itself that remains the same in the welter of change. And just as was the case with Kant, who tried to be scientific without really wading into science, these model-making theologians too are repeating hollow clichés of some trendy philosophers of science. They take various aspects of the "consensus" of those philosophers for truths established by science itself. They are still to learn that false philosophies can cling tightly, and over long periods, to scientific data and conclusions. References to the now defunct ether as the most evident form of matter had been part of intellectual respectability and sophistication throughout the entire lifespan (well over two hundred years) of Newtonian physics.[18]

This fact alone should caution against espousing the slogan that modern physics has disposed of determinism in nature and therefore miracles are possible. The slogan is taken for a self-evident truth in the article on miracles in the recently published *Encyclopedia of Religion*.[19] The

---

[18] In fact as late as 1871 a physicist no less responsible for the eventual overthrow of the ether than James Clerk Maxwell stated that "the ether is *certainly* the largest body of which we have any knowledge" (Italics added). For many similar statements, see my *The Relevance of Physics* (University of Chicago Press, 1966), pp. 79-85.

[19] See note 7 above. To compound the intellectual farce Kelsey also refers to "the findings of psychosomatic medicine" and "the theory underlying the depth psychology of C. G. Jung" as factors casting doubt "on Western materialist determinism" (p. 550). The point at issue is neither materialism nor determinism but ontological continuity which, of course, is a closed book for most theologians nowadays. It is that continuity which is deliberately ignored

modern physics in question is Heisenberg's uncertainty principle. Its entry in such a matter-of-fact way into a theological survey-article indicates that most theologians have already been conditioned to taking it at face value. They hardly suspect that what they eagerly embrace is not physics but philosophy, and indeed a philosophy which destroys the very possibility of a rational defense not only of the facts of miracles but of the very objectivity of the entire Christian Creed. So much in defense of dealing extensively in this essay with three efforts aimed at bolstering the intellectual status of miracles on the basis of that principle. In none of those three cases was good familiarity with physics united with thorough training in theology, and in all three case the philosophical poverty is glaring.

We have indeed come full circle. Until recently the fashionability of Kant's philosophy was the proof that miracles are not possible, whereas today the fashionability of an equally false philosophy, parading in the garb of modern physics, is taken for a proof that miracles are possible. In such an atmosphere it would have been useless to warn that philosophy perishes with the physics it takes for philosophy. In such an atmosphere it cannot be stated bluntly enough and with no apologies that exegetes like Dibelius and Bultmann, to say nothing of many lesser figures, rested their attack on biblical miracles not so much on biblical hermeneutics as on their philosophical epistemology underlying it. Thus in a lecture given on January 27, 1988 in New York's St. Peter's Lutheran Church, Cardinal Ratzinger came most

in the Copenhagen interpretation of quantum mechanics. See note 10 to Ch. 2.

cautiously to that elementary point, in obvious realization that few if any among the many biblical scholars in his audience had ever taken stock of the epistemology hidden in their hermeneutics. While they would not stir on hearing a philosophical argument, they would look up on learning about a new dissertation from the august Theological Faculty of the University of Basel. A dissertation written in German seems to be enough nowadays to assure the aura of scholarship and originality, though it may contain merely the obvious, namely that Dibelius and Bultmann were up to their ears in Kantian philosophy.

Kantian philosophy, let it be recalled, gave a powerful boost to the apriorism in which Newtonian science was partly born. This is particularly true about the famed inverse square law of gravitation, the epitome of scientific law taken over many generations on behalf of the claim that nature is what it is and cannot be anything else, that is, not even the Creator can come up with a different universe. While philosophers and theologians became willing worshipers of the "laws" of nature, some genuinely believing physicists did not forget that the possibility of miracles follows from admitting the reality of the Creator.[20]

To pay serious attention to those scientists would have been the elementary duty of theologians who, because of their non-scientific training, must take much of science "on faith." There were enough believing Christians among German scientists around the mid-century to refute by their very existence Bultmann's

---

[20] One of them, George Gabriel Stokes, is quoted in ch. 1 to that effect.

INTRODUCTION 13

claim that "it is impossible to use electric light and the wireless . . . and at the same time to believe in the miraculous world of the New Testament."[21] He well exemplified those much publicized theologians who discoursed on science with no proper qualifications in it, a topic which, if properly researched, might create a stunned silence in the theological town and a well justified howling outside.

In Bultmann's case too, pretentiousness had its unmasking in the ignorance which in part supported it. One wonders what was his familiarity with Galvani, Volta, Oersted, Ohm, Ampère, Faraday, Weber, Maxwell, and Marconi—who did not discard the world of the New Testament in order to make possible a world bathed in electric light and radio waves. Had Bultmann cited as counterevidence Edison, as voluble representative of village-atheists, he would have at most been entitled to say that as far as leading electricians go the matter is a draw. Consequently it would have been Bultmann's duty to study the reasons, patently non-scientific, why some scientists believed while some did not.

Contrary to the claim of the Anglican bishop, John A. T. Robinson, author of *Honest to God*, nothing about those reasons has changed through "the exploration of the last recesses of the cosmos by radio-telescope if not by rocketry."[22] In fact he himself acknowledged (and forgot in almost the same breath) that not a logical but a

---

[21] R. Bultmann, "New Testament and Mythology," in *Kerygma and Myth: A Theological Debate*, ed. H. W. Bartsch, tr. R. H. Fuller (London: S. P. C. K., 1957), p. 5.

[22] J. A. T. Robinson, *Honest to God* (London: SCM Press, 1963), p. 13.

merely psychological blow had been delivered by modern science and technology to the belief that there is still room for God in the universe.[23] But since the good bishop did not wish to enter the waters of metaphysics that lie beyond the waters of mere physics, it could not dawn on him that science has nothing to do with his basic reason for rejecting miracles. That reason relates to the bishop's rejection of the central dogma of the biblical revelation that comes to a head in the New Testament. It reveals a God who wants such a personal relation with man as to enter human history to the point of becoming man Himself and for only one reason: To provide the only means whereby man's sins can be atoned in the only way that does justice to God's holiness and copes with sinful man's inability to redeem himself.

To accept or reject a personal God has nothing to do with the differential equations of physics or with any of the marvelous tools and toys physicists regale mankind with. Belief in a personal God implies, of course, much more than sound philosophy. It demands above all an "honest-to-God" self-scrutiny along all the parameters (not excluding the all-important parameter of the voice of moral conscience) of human reflections. A debunking of the personal God should seem, of course, a most logical means of stifling the remorse of that modern personal conscience which is caught in countless conflicts owing to the well-nigh complete emancipation of morals. There is indeed far more *Christian* honesty in the thirty or so verses of the first chapter of Paul's Letter to the Romans, where moral depravity is singled out as the chief reason for disbelief in the Creator, than in thirty-

---

[23] Ibid., p. 16.

thousand or so words making up *Honest to God*. There stylistic clichés serve as so many red herrings to draw attention away from problems, philosophical and theological. Crucial moral issues are simply ignored there.

Fortunately, within a few months following the publication of *Honest to God*, Bishop Robinson came clean on fundamental moral issues. He did so in three lectures delivered in Liverpool Cathedral on October 31, 1963. In their published form these lectures ended up with advocacy of "honesty in sex," whereby he justified extramarital liaisons provided they "*truthfully express* the degree of personal commitment that is there underneath."[24] Freudian psychoanalysts had long been saying exactly the same. But the ineffectiveness of their couches to eliminate guilt-feeling is just another proof of the superhuman strategy of Christ (or of those who "invented" his New-Testament image) that made Him invariably focus on sin. In doing so Christ helped the mind to refocus on the evidence of nature about creation, the very basis for miracles. Rooted in this connection is a major pattern of Christian theological history which trendy theologians would do well to study in detail: Belief in the personal Creator remains alive in the measure in which sincere adherence is given to Christ who based on his miracles his claim about the greatest of all miracles which is his being strictly one (consubstantial) with the Father.

Here too, as in many other cases, a supernatural fact (the miracle of Christ's being) sustains the ability of the mind to hold fast to the naturally recognizable fact of

---

[24] J. A. T. Robinson, *Christian Morals Today* (Philadelphia: Westminster Press, 1964), p. 45.

creation. The liturgical celebration of Christ's death as the source in which all martyrs' deaths had their origin, may indeed be applied to miracles as well. The miracle of His reality remains the most powerful assurance about all those miracles that, in His very words, would be done by those who believe in Him. Miracles should seem to abound even today except for those who take refuge in bad philosophy of which its present most fashionable kind is steeped in the cult of chance. Only they fail to give a definition of chance which is more satisfactory than the handy use of that word to cover-up one's ignorance. Mere words standing for ignorance have always been the sophisticated excuse for not bowing to the fact of miracles.

CHAPTER ONE

Prior to the Twentieth Century

Illogical "Saint" David

In writing about miracles it is hardly possible not to think first of David Hume. In many quarters he is remembered as the one who had once and for all divested miracles of intellectual respectability. Such a reputation is part of the awe in which Hume as a man of intellect is still held. Yet, as far as intellectual construction goes, Hume himself admitted that his theory of understanding resembled not so much an edifice as a heap of bricks.[1] To his credit he also perceived that his premises provided only for one kind of glue, plain sentiment or mere habit, to make those bricks stick together into some sort of intellectual framework. About such an outcome Hume was both very despondent in his truly philosophical moments and also very outspoken. Already his first major philosophical work, *The Treatise on Human Nature*, contains the unabashed declaration: "Reason is, or ought only to be, the slave of passions."[2] That a leader of the Enlightenment did not rather speak of the enslavement of passions to reason may tell some-

---

[1] D. Hume, *Treatise of Human Nature*, Bk. I, Pt. 4, sec. 2, (Everyman's Library edition; London: J. M. Dent, 1940), vol. 1, p. 200.

[2] Ibid., Bk. II, Pt. III, sec. 3, vol. 2, p. 127.

thing of the true nature of the light generated by that much glorified movement. To be sure, by "passions" a dignified, urbane comportment was meant by Hume, the "saint" David to his fellow residents in Edinburgh's New Town. Quiet life was a foremost existential commodity in Hume's eyes. All his philosophy was meant to be a shield against harsh, disturbing truths, especially the ones that bespeak of man's subordination to transcendental dimensions. He correctly perceived that none of those dimensions was a potential threat to a tensionless lifestyle if the idea of God was a matter of mere wishful thinking.

Hume's relentless effort to justify intellectually a Weltanschauung free of transcendental constraints was in part a reaction to zealous Calvinist preachment in the Edinburgh of his youth about God's wrath on those He had positively predestined to hell. Another source was Hume's own personal make-up in which as shown by his classic portraits, Epicurean traits dominated. Epicurus, who figures prominently in the closing section of Hume's *Enquiry*, was certainly Hume's model in seeking the intellectual grounds for a peace of mind which consisted in being left alone by the gods.[3]

---

[3] And with a vengeance whose relevance for Hume's critique of miracles is very relevant though often overlooked. Hume's Epicurus is not the one who actually based the formation of everything on the chance swerving of atoms, but a representative of the strict cause-effect method based on empiricism! Assuming that his reader will overlook his manhandling of the record, Hume then expects his reader to admit that, on the basis of a non-empiricist or an a priori philosophy, the power of creating matter out of nothing can be ascribed not only to the Creator but to any mind, as if the maxim *ex nihilo nihil fit* were not impiety itself. Having gone

Nothing has, of course, ever been so much a threat to that ideal of splendid isolation as the One for whom the "Hound of Heaven" is still the psychologically most expressive name.[4] Awareness of Him has three main sources of which one, moral consciousness or the sense of the holy, affected Hume little if at all. In a sense, however civilized, he was a counter image of a Bunyan, a Wesley, let alone of John Henry Newman.[5] He did not pretend indifference toward the two other sources, one philosophical, the other historical. This is not to suggest that his style showed emotional overtones as he went about dissecting the classical proofs of the existence of God: the cosmological and the teleological. He skillfully played the role of uninterested bystander intent only on incontrovertible verities. He could not, however, mask his sarcastic contentment as he completed his picking

---

through this double somersault, Hume's reader is supposed to be dazed enough not to remember that Epicurus was a chief proponent of that maxim which makes creation and miracles impossible by the same stroke. Hume does all this in a seven-line footnote to his *An Enquiry Concerning Human Understanding* (Chicago: Henry Regnery, 1956), p. 172.

[4] A most appropriate caption to Ps. 138 (139) in *The Psalms. A New Translation* (London: Collins, 1963), p. 237.

[5] Since miracles serve primarily the moral order, Hume could have only been infuriated by Newman's relentless emphasis on moral consciousness as a conclusive proof of the existence of a Holy God. Hume would have, of course, reacted with a paroxysm of rage to Newman's no less emphatic insistence that the frequent occurrence of miracles within the Roman Catholic Church was a chief mark of its divine origin. See on this, ch. 3. in my *Newman's Challenge* (forthcoming).

apart the cosmological argument with a celebration of the idea of aborted, incoherent, botched-up universes.[6]

That such an outcome was destructive of the notion of universally valid laws did not seem to bother him. Nor did he seem to be mindful of the fact that years earlier his attack on miracles[7] assumed the notion of laws of nature that were immutable because they were taken to be neccessarily true. He also failed to come to grips with the fact that on the basis of the bare inductionism he advocated one could never have established the existence of laws, let alone of laws that were necessarily true and therefore could not be except what they were. An inductionism severed from metaphysics could not yield even that completeness which was meant by universally and permanently valid laws. Most importantly, Hume did not offer satisfactory explanation of the role he accorded in his philosophy to the recognition of facts. Insofar as he was a sensationist or empiricist philosopher he had to grant equal credibility to the recognition of any fact, usual or unusual. That recognition had to be certain, of the philosophy built on it was to give assurance of certainty. But as far as he was a genuinely Humean philosopher, who subordinated reason to sentiments, he had to part with his professed impartiality vis-à-vis any kind of fact, usual or unusual.

Partiality for some facts, which meant distrust for other facts, invited uncertainty about all facts. This is why when arguing against miracles Hume switched

---

[6] D. Hume, *Dialogues Concerning Natural Religion*, ed. N. K. Smith (Edinburgh: Thomas Nelson and Sons, 1947), Pt. V, pp. 167-169.

[7] "Of Miracles," in D. Hume, *An Enquiry Concerning Human Understanding*, p. 121.

grounds.[8] From a mere probability argument against miracles (the trustworthy witnessing of regular recurrences far outweighed that of exceptional events) he went on dismissing entirely the credibility of witnesses (whatever their number, learnedness, and integrity) on behalf of exceptional or "miraculous" events. Behind such a tactical shift there had to be a fundamental consideration at play. Hume gave a glimpse of it as in the same context he declared, "If the spirit of religion join itself to the love of wonder, there is an end of common sense."[9] Hume's philosophy in general and his arguments against miracles in particular, are pivoted on the meaning he gave to common sense

In Hume's century common sense had many champions, the first of them being the Jesuit Claude Buffier, today almost completely forgotten. They all believed that common sense was the best assurance for certainty about the existence of external reality, that is, of objective facts, things, and events.[10] That Father Buffier's contention had no less an admirer than Voltaire[11] was not

---

[8] As pointedly noted by R. Swinburne, *The Concept of Miracle* (London: Macmillan, 1970), pp. 16-17.

[9] D. Hume, *An Enquiry Concerning Human Understanding*, p. 117.

[10] Buffier wanted to ward off the solipsism which the Cartesian "inner" sense ushered in, by postulating a "common" sense. That he assigned as its primary function the recognition of the external world shows that his intentions were right. But insofar as it was a "sense" anterior to reason, it could logically become, through its articulation by Reid and Lamennais, the kind of subjective intuition whereby one postulates, almost irrationally, the existence of an external world in order that it may be known. See on this development E. Gilson, *Thomist Realism and the Critique of Knowledge*, trans. M A. Wauck (San Francisco: Ignatius Press, 1984), pp. 33-37.

[11] Ibid., p. 33.

without a common though least noted instructiveness. The praises of an unsound *philosophe* were not the clue toward the best philosophical meaning of which an expression, be it common sense, was susceptible. Insofar as it became equated with mere common opinion, it had no chance in securing, and certainly not among philosophers, a certainty about the reality of objective facts. That a commonly shared opinion is a most variable commodity was implied in the phrase "climate of opinion," made popular early in this century by Whitehead, who borrowed it from a "seventeenth-century writer" (Joseph Glanvill) without naming him.[12]

A recall of the times which witnessed the coining of that phrase should not appear useless for Christian theologians. They all, of course, know that defense of miracles has from those times on been an increasingly uphill battle against the climate of opinion taken for common sense. But perhaps not all of them are aware of the unsoundness of a defense of miracles that seeks an ally in the successive climates of opinion, philosophical and scientific. As will be clear later, the present status of that defense provides much for a new chapter in a now old story. That there will be no end to it may be surmised by those mindful of a biblical phrase with a skeptical touch, "There is nothing new under the sun" (Ecclesiastes 1:9), In respect to controversies about miracles, history turns out to be once more the past written in that present tense in which grammarians have long ago recognized the beckoning of the future.

---

[12] A. N. Whitehead, *Science and the Modern World* (1925; New York: The New American Library, 1959), p. 11.

Inverting the inverse square law

But there is an additional reason in an essay on miracles and physics to go back to the past in order to understand the present and be prepared at least for the immediate future. The reason relates to the measure in which physical science contributed to the formation of climate of thought—as difficult to escape as the air one breathes—ever since the days of Glanvill. Those days saw the rise of Newtonian science with all its dazzling successes that opened unsuspected vistas in man's understanding of nature. No less dazzling, though in an opposite or blinding sense, were some philosophical presuppositions grafted on that science by men of science, Newton included, dabbling in philosophy, and by philosophers with little if any expertise in science.

The chief and strictly scientific lure in that dubious game was the inverse square law. Had the seventeenth century not been the age of scientific genius, it might have witnessed serious interest in its own immediate intellectual past. As that past was in good part a matter of printed record, disregard for it could not be excused with a reference to the difficulty of gathering manuscript material, a task difficult even in this age of instant copies and tele-copying. Even a cursory reading of the printed record in question, say of the works of Kepler, Horrocks, and Hooke, would have shown that the inverse square law was not a generalized statement derived from individual observations or experiments. Such an interpretation of the provenance of the inverse square law would have fitted only the empiricist-inductionist straitjacket tailored by Bacon for science as its foolproof method. For that garment, which only some foolhardy amateurs cared to don, no scientific genius of the century of genius had any use. Certainly not Newton. But Newton was also a genius in that he was most unwilling to credit other geniuses. Because of his jealousy of Hooke, Newton failed to give credit to Hooke's ideas on the inverse square law of gravitation.[13] To anyone familiar with Leibniz's work, equally suspect would appear Newton's claims about his having been the sole discoverer of infinitesimal calculus.

Newton certainly belied his own overbearing self-centeredness when he voiced a twelfth-century statement (resurrected in the seventeenth-century debate concerning old and new learning) that the moderns saw farther because they were sitting on the shoulders of giant forebears.[14] Newton certainly had such a giant to help him see the inverse square law loom large on his mental horizon. The giant was none other than Kepler whose three laws of planetary motion were mediated to Newton through a little-appreciated mid-seventeenth-century English astronomer, Jeremiah Horrocks. Those three laws and Huygens' law of centrifugal force could easily be combined in such a way as to yield the inverse square law.

Whether Newton had performed early enough that elementary algebraic operation is disputed.[15] But he did

---

[13] Although R. S. Westfall's sympathies lie with Newton, his article on Hooke in the *Dictionary of Scientific Biography* (New York: Charles Scribner's Sons, 1972), Vol. VI, pp. 481-88, sheds enough light on this matter.

[14] See my Fremantle Lectures, *The Origin of Science and the Science of its Origin* (Edinburgh: Scottish Academic Press, 1978), p. 16.

[15] See R. S. Westfall, *Never at Rest: A Biography of Isaac Newton* (New York: Cambridge University Press, 1980), pp. 387-88.

not need to do so in order to convince himself about the validity of the inverse square law. As one with keen interest in optics, Newton was certainly familiar with Kepler's explicit statements on the decrease of light intensity with the square of distance from a point source.[16] Nor could Newton be ignorant of the fact that Kepler's own certainty about that law of optics was not based at all on observations. Reliable photometry was still two centuries away. Kepler's certainty rested on an a priori philosophical assumption about nature. According to that assumption space was homogeneous. The spreading of the intensity of any physical effect—optical, thermal, or gravitational—in such a space could only follow the inverse square law. Those aware of the influence of the Cambridge Neoplatonists on young Newton, with their markedly a priori speculations, and of old Newton's divinization of homogeneous Euclidean space as God's sensorium,[17] will easily perceive the irresistible attractiveness which Kepler's train of thought had to have on Newton. The latter had been fully convinced about the inverse square law of gravitation long before he compared the fall of the moon in its orbit with the fall of an object on the earth and before he had

---

[16] Those statements are in Kepler's widely read textbook on optics published in 1604. For relevant passages in English translation, see my *The Paradox of Olbers' Paradox* (New York: Herder and Herder, 1969), p. 33.

[17] The subject has been amply treated in the second half of A. Koyré, *From the Closed World to the Infinite Universe* (1957; New York: Harper and Brothers, 1958).

elegantly derived that law from the notion of a central field of force.

The first important thing to note here is the a priori certainty as the source of Newtonian science pivoted on the inverse square law of gravitation. This source was duly and quickly overlooked as Newtonian science proved ever more successful, but it did not fail to act less potently. Newton could lull himself into believing that he was really a "Newtonian" natural philosopher, starting from facts, experiments, and observations and never from hypotheses or postulates. Few leading men of science have ever indulged in so many a priori hypotheses as the one who boasted: *hypotheses non fingo*.[18]

These historical details about seventeenth-century science will reveal their bearing on our topic as soon as one notes the second important point. It is implied in the first point about the certainty felt on a priori grounds about the inverse square law. This chief and spectacularly fruitful law of Newtonian science could easily create a most portentous illusion about the status of the laws of nature. Was it not tempting to assume that the laws of nature were not only accessible to the human mind on an a priori basis but were also ontologically necessary in the form in which they were unveiled by a priori reasoning? And, if such was the case, could there be any real need for experimentation? More importantly, could there be any need for a Creator if nature necessarily had to be what it appeared to be through that very same a priori reasoning?

---

[18] As shown by A. Koyré in his *Newtonian Studies* (Cambridge: Harvard University Press, 1963).

Such questions about science as well as natural theology could hardly be answered in a clear-cut way in the seventeenth century, a transition from Christian to secularized thought. Ambivalence, hesitation, and confusion about these questions were everywhere in the utterances of Galileo, Descartes, Newton, to say nothing of lesser figures with quite a few divines among them, all eager to explore the new science for the purposes of theology.[19] As to science, the potential pitfalls were in full evidence when Galileo slighted Kepler's discovery of the elliptical orbits of planets. He did so on the patently a priori ground that the heavenly motions had to be perfectly circular, a position which allied Galileo with Aristotle, the cosmologist-scientist, whom he wanted to vanquish above all.[20] Descartes could never warm up to the indispensability of experiments.[21] Had Newton not been the scientific genius he was, his philosophical a priorism (with strong Cartesian touches) would have weakened his attention to observational evidence on more than few occasions.[22] He was not attentive at all to

---

[19] The topic, which has generated a large literature, is treated in the perspective of a realist metaphysics in ch. 6 of my Gifford Lectures, *The Road of Science and the Ways to God* (Chicago: University of Chicago Press, 1978).

[20] Galileo's attitude was all the more reprehensible because Kepler explicitly called his attention to the elliptical orbits as established in his *Astronomia nova de motibus stellae Martis* (1609).

[21] On the one hand Descartes urged all *savants* to communicate their experiments to him as the one who alone could properly interpret them, on the other, he sought refuge in the alleged difference between the real world and a world of true laws when confronted with experimental evidence at glaring variance with his theories.

[22] Newton did not, of course, wish to be known as a Cartesian. In

a very specific question raised by that evidence. If the latter was the ultimate truth about the law of nature, what was the true heuristic value of the a priori approach?

This question might have been definitively answered in the seventeenth century had its scientific geniuses not represented a transition between Christian and secular thought. The minds of Galileo, Descartes, and Newton were not sufficiently Christian to perceive the pseudo-deity (be it pantheism or materialism or deism) lurking behind a priori thinking. They were, however, sufficiently Christian to give no serious consideration to a world without a Creator. Belief in creation thoroughly conditioned their minds to think of nature as suffused by reason and law even in that sublunary realm which was a realm of disorder for pagan minds such as Plato and Aristotle, whatever their intention to celebrate cosmic rationality. But if one was to consider in all its consequences the denial of an infinitely rational Creator, could the inference be avoided that in such a case no rationality whatever could appear in nature?

This inference was not broadly recognized even when in the nineteenth century the secularization of Western thought began to unfold its full logic. Even such a master logician as John Stuart Mill failed to recognize the full implications of his own speculation about other worlds where two and two would not necessarily make four.[23] For in that case there beckoned the specter of

---

his later years he spent much time erasing the name of Descartes from the manuscript notes he took as a young scientist.

[23] For details and documentation, see my *Road of Science and the Ways to God*, pp. 390-391.

absolute contingency. Against that nihilistic possibility Mill's 'god' (half good, half evil) was hardly a logical shield. Mill was, however, logical to the point of admitting that Hume's notion of the invariability of nature's laws as a refutation of miracles rested on the presupposition that God does not exist. For as Mill put it, a "miracle is a new effect supposed to be produced by the introduction of a new cause . . .; of the adequacy of that cause, if present, there can be no doubt."[24] Even less recognition was given during the nineteenth century to the fact that absolute contingency was but a replay of the occasionalism advocated by al-Ashari and Ockham, both of whom refused to grant the Creator the ability to create a consistently acting nature.

At any rate, was the mere semblance of order in nature that alone could be had within in absolute contingency or occasionalism different from the order admissible in radical positivism? Were not in the latter the laws of nature mere commodious devices created by the mind for its own convenience to deal with facts?[25] Scientists who endorsed that theory did not do so to the logical extent of advocating a closure of laboratories. This conflict of that unsound theory with their sound practice provided no answer to the question as to why there was order in nature, that is, why nature obeyed specific laws.

---

[24] J. S. Mill, *A System of Logic, Ratiocinative and Inductive*, III, xxv, 2 in *The Collected Works of John Stuart Mill* (University of Toronto Press, 1970), vol. 7, p. 623.

[25] Commodism received its most concise formulation when Poincaré declared that "one geometry cannot be more true than another; it can only be more convenient." *Science and Hypothesis* (New York: Dover, 1952), p. 50. The French original was published in 1902.

Refusal to anchor that order in the Creator could but leave one with the fearful prospect of a radically random state of affairs. There stones would not regularly fall, but just as likely hang in mid-air or take off unexpectedly in any direction. There it would be most unlikely that the hatching of a chicken egg would yield a chick. There a flower would perpetuate its own kind only as an exceptional case. In other words, in a world severed from its Creator, lawfulness would be the miracle, that is, a most unexpected event.

No miracles, no things

The foregoing examples are a mere paraphrase of the most incisive pages written on the laws of nature. The time, 1908, was the highwater mark of the worst misrepresentation of science, which is on hand whenevever science is cast in the mold of scientistic ideology. The book, not surprisingly, had *Orthodoxy* for its title. Its author, Gilbert Keith Chesterton, was not a scientist, not even a philosopher of science.[26] But he was certainly a thinker most independent of the climate of thought of his time if there ever was such a thinker. No such independence was evident in the geniuses of the seventeenth century, with Pascal being a major exception. A famous remark of his portrayed Descartes as the one who really had no need of God.[27] Pascal could just as well have

---

[26] Yet, the relevant chapter, "The Ethics of Elfland," of *Orthodoxy* was in part reprinted in 1957 in a book, *Great Essays in Science*, put together by Martin Gardner, associate editor of *Scientific American*. For details and discussion, see my *Chesterton, A Seer of Science* (Urbana: University of Illinois Press, 1986), pp. 12-16.

[27] *Pascal's Pensées*, trans. W. F. Trotter, with an introduction by T.

stated the same about Galileo, who could exalt the human mind and debase the divine mind in the same breath. When he stated that the human mind was the greatest marvel of creation, Galileo also equated man's knowledge of geometry with the Creator's knowledge of it.[28] From there it was but a step to deriving nature's geometrical structure from the mind. Such a nature soon was seen to be in no need either of mind or of God. The perception was achieved just as decade or so before science came of age through Newton's *Principia*. In that perception as offered by Spinoza, nature and God were made identical which simply excluded the possibility of miracles.[29] Much less noted was the most devastating consequence of the Spinozian position. It consisted in Spinoza's thorough perplexity about the existence of concrete, specific, limited things making up nature and providing science with its subject matter.[30]

S. Eliot (New York: E. P. Dutton, 1958), p. 23.

[28] Galileo did so with a distinct touch of apriorism. See the concluding pages of the "First Day" of his *Dialogue Concerning the Two Chief World Systems*, trans. Stillman Drake (Berkeley: University of California Press, 1962), pp. 102-104.

[29] "We may conclude," states Spinoza in ch. vi, "Of Miracles," in his *Theologico-political Treatise*, "that we cannot gain knowledge of the existence and providence of God by means of miracles, but that we can far better infer them from the fixed and immutable order of nature. By miracles, I here mean an event which surpasses, or is thought to surpass, human comprehension: for in so far as it is supposed to destroy or interrupt the order of nature or her laws, it not only can give us no knowledge of God, but, contrariwise, takes away that which we naturally have, and makes up doubt of God and everything else." See *Spinoza's Works* (New York: Dover, 1951), vol. 1, pp. 86-87.

[30] Spinoza's attention to this problem arising from his philosophy

To be sure, few at that time, and certainly not the scientists, became Spinozians. But whereas there was no pressing need for following Spinoza, the question of miracles imposed itself in the measure in which the laws of nature began to appear as subtly ultimate entities. Undoubtedly, Newton was sincere as he assigned to God's direct action certain arrangements in the physical realm for which his science contained no answer. One such arrangement was, according to Newton, the separation of fiery from cold matter, or the separation of celestial bodies into stars and planets.[31] His other example, the periodic intervention of God to secure the stability of the solar system, made a better known intellectual history through Laplace's scientific solution to that problem.[32] It was the history of holes in which

---

was called by E. W. von Tschirnhausen, a gentleman-philosopher from Heidelberg, in 1676. All that Spinoza offered in reply was that he hoped to put the matter "in due order," an impossible project insofar as it was to reduce the order embodied in the specific varieties of things to an order conceived a priori.

[31] A consequence of this was, according to Newton, the sun's uniqueness as a heat-and-light-giving body. Other features of the solar system, for which Newton gave credit to a direct intervention by God, were the proper adjustment of the distances, masses, and velocities of planets, their rotation on their axes, the measure of the inclination of the earth's axis, the exact amount of each planet's angular momentum, and the orbiting of all planets in the same direction and in the same plane. For further details, see my *Planets and Planetarians: A History of Theories or the Origin of Planetary Systems* (Edinburgh: Scottish Academic Press; New York: John Wiley, 1977), pp. 70-73.

[32] It was that solution that prompted Laplace's boastful remark to Napoleon: "Je n'ai pas besoin de cette hypothèse," that is, an intervention by God.

divines, ready to jump on the bandwagon of science without seriously studying it and unwilling to probe an essentially philosophical question to its very bottom, were eagerly locating Almighty God as one busy doing what science could not do for the time being.

Many divines, in fact, lost their faith in miracles as they saw those holes being filled up with the relentless progress of science. They turned to an increasingly radical reinterpretation of biblical miracles culminating in the exegesis of Bultmann and in the philosophy of Ernest Bloch.[33] They might have saved themselves from this intellectual and spiritual debacle had they pondered Newton's own position vis-à-vis miracles. Newton's willingness to admit the reality of biblical miracles alone never cut ice with rationalists. It was quite possibly a tactic on Newton's part to cover up his Unitarianism, which, if discovered, would have cost him the Lucasian chair in Cambridge and, later, the Directorship of the Mint. Unbelievers could, of course, be but reassured by Newton's categorical denial of Christian miracles postdating New Testament times.[34] Clearly, Newton believed

---

[33] According to Bloch, miracles are a mythical projection by man into a still unknown future state in which mankind will reach its "divine" form through the purely natural mechanism of biological evolution. Bloch, of course, merely reiterated, with respect to miracles, ideas that had been set forth by Henri Bergson and Samuel Alexander under such labels as *élan vital* and *nisus*.

[34] Here Newton merely followed none other than Robert Boyle, who wished to be known as the chief "Christian virtuoso" of the new mechanistic science and who claimed to rise from the mechanistic contrivances constituting nature to the "seraphic love" of God. Boyle's dismissal of post-biblical miracles as being unworthy of God, the clockmaker, is a perfect example of the vengeance which

less in Christianity than he should have and believed more than a Christian should in the laws of science and nature. One wonders whether Newton had ever as much as suspected the miracle of creation in the beginning lurked behind all laws of nature, and their totality, or the miracle of a specific nature stable in its orderliness. For only with an eye on that miracle can the possibility of miracle be raised meaningfully.

The miracle of Creation

The miracle of creation in the beginning implies, of course, the Creator's sovereign freedom to create or not to create. No less importantly, his creative freedom is divine also in the sense that the actual universality of things created by Him is only one of an infinite number of possibilities at His disposal. Such a Creator is not contradicting the rationality and consistency of His creation if it includes an intellectual and moral realm which the physical realm is to support and serve. With such a notion, and only with such a notion of Creator and creation in focus, it was natural to say, as did George G. Stokes, a prominent physicist of the late nineteenth-century, "Admit the existence of a God, of a personal God, and the possibility of miracle follows at once."[35] The contrast could not have been greater with Voltaire's often quoted utterance that "to suppose that

---

one's lack of sound philosophy can take both on one's theology as well as on one's broader interpretation of science. For further details, see my *The Road of Science*, p. 89.

[35] G. G. Stokes, *Natural Theology* (London: Adam and Charles Black, 1891), p. 24. This book is the text of the Gifford Lectures, delivered by Stokes at the University of Edinburgh in 1891.

God will work miracles is to insult Him with impunity."[36] The ground for that insult was, according to Voltaire, that a miracle meant the inability on God's part to accomplish any particular end by immutable laws. The fallacy of Voltaire's reasoning lay in his own inability to see in the realm of existence anything but a mere clockwork in which there was no room for free beings. Of human freedom, a most immediately evident factual experience, he could speak only with the gravest perplexity.[37]

Voltaire was not the first or last worshiper of the laws of nature for whom the freedom of the will was a source of continual nightmare. One wonders what latter-day Voltaires felt on hearing no less a physicist than Arthur H. Compton declare that the evidence on behalf of man's conviction to move his little finger at will was immensely greater than all the evidence on behalf of Newton's laws. From this it followed, according to Compton, that should a conflict arise between our sense of freedom and Newton's physics, it is the latter that needed to be revised.[38] Recognition of man's freedom

---

[36] Voltaire, *Dictionnaire ou la raison par Alphabet*, ed. J. Benda and R. Naves (Paris: Garnier Frères, n. d.), vol. 2, p. 147. The same "concern" for the dignity of God is R. W. Emerson's utterance in the address he gave to the senior class in the Divinity College at Harvard College on July 15, 1838: "To aim to convert a man by miracles is a profanation of the soul." See his *Nature, Addresses, and Lectures* in *Emerson's Complete Works* (Boston: Houghton, Mifflin & Co., 1888-90), vol. 1, p. 131.

[37] As, for instance, in his *Le philosophe ignorant* (1766); see *Oeuvres de Voltaire* (Paris, 1877-85), vol. XXVI, p. 55.

[38] See Compton's Terry Lectures, *The Freedom of Man* (New Haven, Conn.: Yale University Press, 1935), p. 26.

means, of course, the recognition of a moral order which alone is to be served by miracles. That they were never meant for the entertainment of the curious was called to the attention of the Pharisees and Sadducees of all ages by none other than the greatest miracle worker of all times, Jesus of Nazareth.[39]

In 1934, when Compton spoke, Newtonian physics had for some time been superseded. The mathematical tools of the new physics—quantum mechanics—were statistical. They were very different from, indeed irreducible to pure differential equations, which are the backbone of Newtonian physics. Those equations are all translatable into geometry in which the lines or curves representing various parameters are always continuous. (It may be worth recalling that in the *Principia*'s mathematical propositions and demonstrations were invariably given in geometrical figures equivalent to what later became known as differential geometry.) The geometrical continuity implied, in principle at least, the possibility of measuring with perfect accuracy the physical processes described by those figures. It was at that point that an elementary error in reasoning gained currency among physicists who were then readily echoed by philosophers overawed by the success of a field they did not really know. The possibility of perfectly accurate measurements became quickly taken as *the* justification of ontological causality in physical interactions. The inference was equivalent to putting the cart before the horse. Worse,

---

[39] His denunciation of the "evil and faithless age" looking for signs in Matthew 12:39 and 16:4 was certainly sharp, but just as keen was the frustration which Jesus felt over his inability to work even more miracles because of the lack of faith in some people.

that mistaken philosophical maneuver began to function as the *exact* foundation of the ideology of immutable laws of nature, of absolute physical determinism, and of the absurdity of miracles.

Resistance to miracles

Spokesmen of that ideology were a dime a dozen. Ironically, their self-confidence and public acceptance reached its peak just in the decades when the handwriting became increasingly visible on the superb edifice of Newtonian physics. The decades were the ones saddling the turn of the century.[40] The moderately comforting aspect of that intellectual debacle was that some prominent men of science did not lend their voice to a chorus wholly out of tune with science. Henri Poincaré, the foremost mathematical physicist of the time, had for all his agnosticism the presence of mind to warn that "it is with freedom that one demonstrates complete determinism."[41] Philosophers and divines lacking that modicum of common sense were all too numerous. Thus E. Goblot, professor of logic at the Sorbonne, wrote "All induction rests on the confidence we have in determinism. There is therefore in nature neither contingency nor caprice, nor miracle, nor free will; any of these hypotheses ruin our

---

[40] Those decades witnessed more than one prominent physicist celebrating classical physics as the final form of man's knowledge of nature. For details, see my *The Relevance of Physics* (Chicago: University of Chicago Press, 1966), pp. 84 and 90.

[41] "C'est librement qu'on est déterministe." Poincaré noted the obvious fact with consummate conciseness in his article, "Sur la valeur objective des théories physiques," *Revue de métaphysique et de morale* 10 (1902)(: 288.

mental ability to reason about things."[42] The only logical merit in that statement was the juxtaposition of miracles and free will. A most glaring fault from the logical viewpoint was Goblot's reference to confidence, hardly a matter for rigorous logic. Last, but not least, one would ask, was Goblot entitled to kudos—and to royalties—for his book if in terms of his declaration it was a necessary result of his brain mechanism, or more rudely, of his bread and board?

Unintended instructiveness is no less glaring in the declaration of the philosopher of religion, A. Sabatier, "Miracles have no basis in modern philosophy. The method inaugurated by Galileo, Bacon and Descartes gives to our thinking a turn which necessarily excludes it."[43] Such a turn could come about only through a very selective reading of those three and of others with whose names Newtonian science became synonymous. Whatever the inadequacies of their dicta on scientific method, those three certainly wanted no part in an ideology restricted to matter and motion. They would have undoubtedly rejected the declaration of G. Séailles, a chief late-nineteenth-century spokesman of empiricist and scientistic secularism, "By its principles as well as by its conclusions science excludes miracles."[44] The

---

[42] E. Goblot, *Traité de logique* (Paris: F. Alcan, 1918), pp. 313-314.

[43] A. Sabatier, *Esquisse d'une philosophie de la religion d'après la psychologie et l'histoire* (Paris: Fischbacher, 1897), p. 80.

[44] G. Séaillles, *Les affirmations de la conscience moderne* (Paris: A. Colin, 1903), p. 32. Séailles' declaration had, of course, been many times anticipated by freethinkers throughout the nineteenth century. Thus Thomas Jefferson wrote to John Adams on June 20, 1815: "The question before the human race is, whether the God of Nature shall govern the world by His own laws, or whether priests

empiricist Bacon's dismissal of miracles as means never used by God "to convert the heathen," was still balanced by his emphasis on the evidence which an orderly nature brings to its Creator,[45] a position unacceptable to Séailles and his cohorts among empiricists. Séailles could hardly be ignorant of Descartes' often quoted dictum "God performed three miracles: the creation of things out of nothing, the freedom of the will, and the Incarnation."[46] The point, from which the scientistic antagonists of miracles might have most profited from and which they would have most resented, was also already made in Descartes' century and by no less a scientist than Leibniz: "If geometry were as much opposed to our passions and present interests as is ethics, we would contest it and violate it almost as strongly, the demonstrations of Euclid notwithstanding."[47] In blissful disregard of this the

---

and kings shall rule it by fictitious miracles." *The Adams-Jefferson Letters*, ed. L. J. Cappon (Chapel Hill, N. C.: University of North Carolina Press, 1959), vol. 2, p. 445. It escaped Jefferson that precisely by allowing God to perform miracles, priests truly recognized God's laws about nature to be really *His* and not of scientists and philosophers making half-hearted concessions to a God who was as much the prisoner of his own laws as they were to their own preconceived notions.

[45] The position, which is introductory to Bacon's essay "Of Atheism," is certainly characteristic of the heavy drifting of many seventeenth-century Puritans from classic Calvinist positions concerning natural theology.

[46] "Tria mirabilia fecit Dominus: res ex nihilo, liberum arbitrium & Hominem Deum," reads the original in Descartes' youthful "Cogitationes privatae." See *Oeuvres de Descartes*, ed. C. Adam and P. Tannery (Paris: L. Cerf, 1897-1913), vol. 10, p. 218.

[47] *New Essays on the Human Understanding*, trans. A. G. Langley (London: Macmillan, 1896), p. 93 (Bk. 1, ch. ii, 12). Were this to

pseudo-intellectual's sneering at miracles grew into a crescendo in the decades straddling the turn of the century when Christian morals, private and public, which Christian miracles were to support above all, became for the first time a major target of secularism.

---

happen, continues Leibniz' mouthpiece, those demonstrations would be called "dreams" and "full of paralogisms."

CHAPTER TWO

In the Twentieth Century

Miracles and Einstein's relativity
As the twentieth century began to unfold, physics revealed two unsuspected aspects of it. Of those two, the theory of relativity, showed no direct relevance to the question of miracles. It has, however, an important indirect bearing on that question, which is worth recalling in a few words and all the more because it is generally overlooked or simply ignored. The theory of relativity was born out of young Einstein's awe for the intellectual beauty of Maxwell's equations.[1] Since beauty is inseparable from form, it was all too natural for him to be concerned about the distortion of the simple form of those equations as they are applied to a reference system moving with respect to the observer. Einstein's great insight consisted in perceiving that the transformation of those equations from one reference system to another would leave intact their form provided the speed of light is taken for something absolute, independent even of the speed of its source. A quick recall of the fact that the

---

[1] They impressed him as if they were a "revelation," to recall his reminiscences in his "Autobiographical Notes," in *Albert Einstein Philosopher-Scientist*, ed. P. A. Schilpp (1949-51; Harper Torchbook, 1959), vol. 1, p. 33.

speed of sound is never independent of the motion of its source may help one realize the enormity of the departure which the Einsteinian postulate of the absoluteness of the speed of light represents with respect to Newtonian physics. In the latter, which is rightly spoken of as mechanistic physics, the speed of the propagation of any mechanical or physical effect is always a function of the speed of its source. That Einsteinian relativity is based on the unconditionally absolute value of the speed of light may also help one perceive the measure of skullduggery whereby the relativization of all ethical and social values is asserted on the basis of Einstein's relativistic physics. The latter is the most absolutist physical theory ever proposed.[2]

The foregoing considerations relate to the theory of special relativity which, with its uniformly moving reference systems, is a particular case of the theory of general relativity. The latter deals with accelerated frames of reference. Since the most obvious case of acceleration is gravitational motion, it was almost inevitable that Einstein should make an effort to deal with the gravitational interaction of all matter. He did so as he presented in 1917 the last of his memoirs on general relativity. That memoir was a great first in physics in that it contained the first, contradiction-free scientific account of a gravitational universe.[3] The im-

---

[2] As I argued in my article, "The Absolute Beneath the Relative: Reflections on Einstein's Theories," *The Intercollegiate Review* 20 (Spring/Summer 1985), pp. 29-38. Reprinted in my *The Absolute beneath the Relative and Other Essays* (Lanham, Md.: University Press of America and Intercollegiate Studies Institute, 1988).

[3] The price of that success was, of course, a parting with a

portance of this can easily be seen with a recall of the point on which Kant staked his critique of the cosmological argument. The point was that science (the science of Kant's time as he poorly knew it) provided no contradiction-free account of the universe. This is why Kant felt entitled to call the notion of the universe a bastard product of the metaphysical cravings of the intellect and, therefore, unsuitable to serve as the final and crucial jumping board in the intellectual recognition of the existence of the Creator.[4] This objection of Kant continues to command credibility only on the part of those unmindful of Einstein's achievement, which should, however, loom large in the eyes of those hopeful about a genuine harmony between science and miracles. The latter can be part of rational discourse only if the existence of the Creator and of a moral order (inconceivable without Him) are assumed. Einstein's contribution to the scientific grasp of the universe should seem therefore of greatest importance. In fact he perceived late in his life that his cosmology may be an unintended pointer to the One beyond the *totality* of consistently interacting things which is the universe.[5]

---

universe infinite in the Euclidean sense, a universe plagued also by the optical paradox, better known as Olbers' paradox. But since that infinite universe had often been taken during the nineteenth century for a substitute ultimate entity, the net gain for natural theology was enormous. For a more detailed discussion, see my article, "The Intelligent Christian's Guide to Scientific Cosmology," *Faith and Reason* 12 (No. 2, 1986), pp. 124-36 and "Teaching of Transcendence in Physics," *The American Journal of Physics* 55 (1987), pp. 884-88.

[4] On Kant's strategy see ch. 8 in my *The Road of Science*.
[5] He did so in his exchange of letters with M. Solovine in 1950-51.

The quantum jump of non-sequitur

While the indirect support which the theory of relativity brings to miracles remains unexploited, quite a vast literature has arisen about the alleged support which the other main branch of modern physics, quantum theory, allegedly has for miracles.[6] That literature certainly proves the naiveté with which theologians try to cash in on science even when they are not properly trained in it, or appraise it with false philosophical premises. They still have to learn that a wrong starting point can only lead to blind alleys regardless of the subject, be it as lofty as theology or as down-to-earth as physics. In following up philosophical blind alleys, theologians who stake their fate and fortunes on the divinity of the Logos, that alone makes Christian miracles reasonable and meaningful, should view most seriously any misstep in logic, in particular, and philosophy in general. It should seem

---

For various passages in English translation from those letters, see my *Cosmos and Creator* (Edinburgh: Scottish Academic Press, 1980), pp. 52-53.

[6] The contribution of theologians to that literature can be gauged from the lengthy footnote in the best modern Roman Catholic monograph on miracles: L. Monden, *Signs and Wonders: a Study of the Miraculous Element in Religion* (New York: Desclee, 1966). His own comments (pp. 329-330) make it clear that the dozen or so publications listed by him were not such—and this is certainly true of the best of them, F. Selvaggi, "Le leggi statistiche e il miracolo," *La Civilta Cattolica* 101/IV (1950), pp. 45-56 and 202-213—as to make him perceive the core of the issue or, the non sequitur of inferring from the uncertainty of measurements to an ontological incompleteness in natural interactions. The same is true of the equally representative recent Protestant monograph by Colin Brown, *Miracles and the Critical Mind* (Grand Rapids, Mich.: Wm. B. Eerdmans, 1984), pp. 178-79.

most un-Christian to espouse mental somersaults or plain verbal tricks that abound in the literature on the philosophy of quantum mechanics as well as on the demythologization of miracles.[7]

As to the philosophy of quantum mechanics, the pattern for somersaulting was provided by none other than Heisenberg, one of the architects of quantum mechanics and the first to unfold a principal consequence of it. Since its formulation in 1927 that consequence has made intellectual history (not necessarily coextensive with the history of truth) under the label of the principle of indeterminacy or uncertainty principle. A much less misleading label would have been the principle of imprecision. For what Heisenberg found was simply that measurements of physical interactions involving conjugate variables, such as momentum and position, time and energy, will always contain a margin of imprecision, which can be significant on the atomic level. (On the level of ordinary perception or macroscopic level, the quantum mechanical imprecision can be safely ignored because it is many magnitudes smaller than the probable error acceptable for laboratory or industrial practice on that level.) Heisenberg, however, jumped to the conclusion that because of the significance of inevitably impre-

---

[7] Few students of Bultmann's method dare to be so outspoken as L. J. McGinley was in his *Form Criticism of the Synoptic Healing-Narratives* (Woodstock, Md.: Woodstock College Press, 1944) in commenting on Bultmann's demythologization of New Testament miracles: "It is such a mixture of arbitrary statements and detailed analysis, of capricious bias and clever dissection that it leaves the reader overwhelmed and confused" (p. 43).

cise measurements on the atomic level, the principle of causality should be considered as overthrown.[8]

This inference was not so much a brave conceptual quantum jump as a needless *metabasis eis allo genos* or a sheer non sequitur, although not without an important though often overlooked merit. If, indeed, the imprecision in question meant the overthrow of causality, the latter could not be salvaged on the ground that the imprecision in question is wholly negligible on the macroscopic level of ordinary existence and operations. The absence of the ontological factor, causality, in the foundations cannot issue in its presence in a superstructure which is their extension. At any rate, was Heisenberg right in claiming that there was no causality because of the inevitable imprecision of measurements of physical interactions? That question should have been answered in the negative. Instead, it was given an affirmative answer and to the extent as to become a climate of thought.[9]

---

[8] W. Heisenberg, "Uber den anschaulichen Inhalt der quantentheoretischen Kinematik und Mechanik," *Zeitschrift für Physik* 43 (1927): 197: "The invalidity of the law of causality is definitely proved by quantum mechanics." Those who naively believe that great physicists reach momentous philosophical conclusions under the impact of their scientific finds, will find eye-opening the study of P. Forman, "Weimar Culture, Causality, and Quantum Theory, 1918-1927; Adaptation by German Physicists and Mathematicians to a Hostile Intellectual Environment," in *Historical Studies in the Physical Sciences*. Vol. 3 (Philadelphia: University of Pennsylvania Press, 1971), pp. 1-115. Forman shows that Heisenberg was one of a dozen prominent German physicists who had decried causality for several years before 1927.

[9] This climate of thought was memorably capsulized by H. Margenau: "No simple slogan, save 'violation of causal reasoning'

It was largely overlooked that Heisenberg's principle states only the inevitable imprecision of measurements on the atomic level. From that principle one can proceed only by an elementary disregard of logic to the inference that *an interaction that cannot be measured exactly, cannot take place exactly.*[10] The fallacy of that inference consists in the two different meanings given in it to the word *exactly*. In the first case it has a purely operational meaning, whereas in the second case the meaning is decidedly ontological. The inference therefore belongs in the class of plain non sequiturs that, as a rule, are severely structured in better-grade courses on introductory logic.

The alleged demise of ontological causality should have called for a general sounding of alarms. For that demise could mean but the opening of a chaotic abyss with neither a bottom nor safe perimeters limiting its extent. A recognition of this, coupled with a consistent attention to it, could not be expected either on the part of prominent physicists or on the part of those in excessive awe of their mental prowess. Einstein's admission that the man of science is a poor philosopher has much more to it than meets the eye. He himself failed to suspect this as he lead a very small group of physicists who refused

---

was deemed sufficiently dramatic to describe the revolutionary qualities of the new knowledge." *The Nature of Physical Reality* (New York: McGraw Hill, 1950), p. 418.

[10] This is my rephrasing of Turner's remark, quoted later, and a fair summary of a chief contention in my article, "Chance or Reality: Interaction in Nature versus Measurement in Physics," first published in 1981 and reprinted in my *Chance or Reality and Other Essays* (Lanham, Md.: University Press of America and Intercollegiate Studies Institute, 1986), pp. 1-21.

to accept the counter-causal twist which Heisenberg gave to his principle and which later became the cornerstone of the Copenhagen philosophy of quantum mechanics with Niels Bohr and Max Born as its chief articulators. Einstein never came to realize fully that his disagreement with those two was not so much about causality, which he too equated with the possibility of perfectly precise measurements,[11] but with the ontology implied in causality, physical or other. It was left for W. Pauli, another prominent physicist, to call to this point the attention of Born who grew as much dismayed by his inability to convince Einstein as by the cooling of Einstein's feelings toward him. But Born could hardly derive much enlightenment from Pauli's scornful remark that Einstein's concern for physical reality was not worth more than the medievals' debate about the number of angels that could be accommodated on a pinhead.[12]

Desperate rescue operation

The inability to articulate the question of ontology underlying the debate on the status of causality in the perspective of quantum mechanics took monumental proportion in some lengthy essays of Planck on causality,

---

[11] This is why Einstein looked for the success of hidden variable theories and this is why he kept speculating about a thought experiment which would show the possibility of measuring with perfect accuracy interactions involving conjugate variables.

[12] For details and documentation, see my *Chance and Reality*, p. 10. Pauli's scoffing at the ontological question was not untypical of the attitude of many of his fellow physicists. Relatively unimportant is their readiness to perpetuate that allegedly medieval preoccupation whose first written appearance does not antedate the fifteenth century, well known for its scorn for anything medieval.

world order, and freedom.[13] Ontology and the consequent distinction of it from the merely operational level could hardly emerge on the mental horizon of Planck, a professed Neo-Kantian. For him, causality was a mental category which did not depend at all on the observation of the external world. Within the iron grip of that category were, according to Planck, all events, including all mental operations, even those of the greatest geniuses. Consequently, the freedom of the will as a mental decision could be but a practical convenience resulting from the fact that our introspection did not permit a fully objective, that is, completely accurate evaluation of our motivations. It was that practical impossibility that, according to Planck, made even Laplace's superior spirit a free agent. As to the Supreme Wisdom or God, Planck refused to discuss whether He too was free only in that practical sense, or whether He was not free at all because His nature implied a perfectly accurate introspection.

The word accurate is worth noting because the possibility of accurate, that is, quantitatively exact measurements was an integral part of Planck's notion of causality. He borrowed it from the physicists' world in which he lived and worked. There the notion of causality had been as widely based on the notion of exact measurements prior to the advent of quantum mechanics as was the denial of causality following the overthrow by quantum mechanics of their practical possibility. In a broader cultural consciousness, the foregoing shift

---

[13] For further details, see my article, "The Impasse of Planck's Epistemology," in *Philosophia* (Athens), 15-16 (1985-86): 143-165. Reprinted in my *The Absolute beneath the Relative and Other Essays*, pp. 18-42.

appeared as a departure from a deterministic notion of nature to a non-deterministic one. The inference that thereby belief in the freedom of the will received a scientific approval was quickly made, and by no less a scientist than Eddington.[14] Much less attention was given to his reconsideration of the matter, a few years later, in 1939 to be specific, when he declared that his earlier arguments on behalf of the freedom of the will on the basis of the uncertainty principle were wholly mistaken.[15] References to a new "scientifically" approved approach to the freedom of the will kept popping up in the philosophical and theological literature, a story that may be worthy of detailed documentation. Philosophers and theologians may not be less inclined to learn from the errors of the past than are politicians and scientists.

That the origins of scientifically-coated rescue operations on behalf of the freedom of the will antedate the advent of quantum mechanics shows that theologians can be quite naive in trying to cash in on some glittering straws in the wind. The simile may seem all the more appropriate because it relates to the development of statistical methods in gas theory during the closing decades of the nineteenth century. That this development was often appraised well outside theological circles as a departure from the deterministic world view is, of course, true, but this is not necessarily a guarantee of reliability. As a matter of fact, the statistical gas theory was based on a strictly deterministic application of the

---

[14] Eddington did so at Cornell University. See his *The New Pathways of Science* (Cambridge University Press, 1934), p. 88.

[15] He in fact strictured his earlier suggestion as nonsensical in his *The Philosophy of Physical Science* (London: Macmillan, 1939), p. 182.

Newtonian laws of motion about the collision of gas molecules taken for perfectly elastic and spherical bodies. In such a situation, the initial conditions determine with complete accuracy any subsequent state, however far removed from the initial state. A rigorous interpretation of gas theory did not therefore justify the inference that most out-of-the-ordinary configurations were only most improbable but not inherently impossible. They were impossible in the measure in which the initial conditions were ordinary, that is, fairly symmetrical configurations or not. When some early twentieth-century defenders of miracles reported that "the old rigid system of the laws of nature is being broken up by modern science,"[16] they were very far from reliable scientific grounds which are always very different from current fashionable appraisals of the latest in science. The same is true of some scientists who tried to discredit miracles by calculating the enormous improbabilities of deviations from the ordinary course of nature. The figure 10^{100} given by J. Perrin, a French Nobel-laureate physicist, to illustrate the improbability of a tile to deviate from its vertical fall[17] may impress even the layman by its being incomparably larger than all the atoms in the universe and all the microseconds that have elapsed since its expansion got under way sixteen billion years ago.

---

[15] As, for instance, J. T. Driscoll in his article "Miracle" in *The Catholic Encyclopedia* (New York: The Gilmary Society, 1913), vol. 10, p. 341, who in turn took that phrase from the October 1908 issue of *The Biblical World*.

[17] The figure in question is about 20 magnitudes larger than the total number of atoms in the expanding universe. When a calculation involves such unimaginably large numbers, even a wide margin of error fails to impair its instructiveness.

The super-astronomical improbability of this happening does, however, in no way weaken the certainty of that outcome, provided the initial conditions are in exact conformity with it. But about those initial conditions the scientist could only admit his ignorance, although he should have kept in mind that it was that very ignorance that prompted him to calculate mere averages. Since the latter were useful only for the gaseous state, in which the motion of molecules represents an extreme case, the application to miracles was in fact tantamount to a specious blowing of mere hot air, worthy only of less than average intellects.

Miracles and nominalism

The grafting of scientific respectability on miracles had a far greater appeal with the advent of quantum mechanics as it was taken to be equivalent to the breakdown of strict physical determinism. Here again a detailed account of what actually happened is still to be written. That there was an early rush of theologians to a terrain which, as will be clear later, was a ground where angels would have feared to tread, may be gathered from a book of Bernard Bavink, published in 1933 and immediately translated from German into English under the title, *Science and God*.[18] The book was the substance of lectures which Bavink had given in various parts of Germany on science and religion. A graduate of the University of Göttingen, where he majored in physics, Bavink had a deep interest in theological questions. This was almost

---

[18] B. Bavink, *Science and God*, trans. H. Stafford Hatfield (London: G. Bell & Sons, 1933).

natural on the part of one who had among his paternal forebears Dutch Mennonites and was converted to Lutheranism by his wife, the daughter of a pastor. By the early 1930s Bavink had for some years been looked upon as a leading Christian interpreter of the relation between science and religion. This was due to the half a dozen editions, between 1913 and 1929, of his magnum opus that appeared in English translation in 1932 under the title, *The Anatomy of Modern Science*.[19] That miracles and science are not discussed in those editions (and in that translation) is an indication of the fresh interest in that topic by the uncertainty principle, still a novelty around 1930 or so.

In recalling Perrin's calculation of the enormous improbability of a macroscopic object, such as a tile, from deviating randomly from its vertical fall, Bavink noted that miracles, such as Peter's walking on the water, were macroscopic events where the laws of classical mechanics were, with their strict determinism, invariably valid. Such was the immediate background for Bavink's warning. "The theological world cannot be too strongly warned against attempting to make capital in this way of the new discoveries."[20]

By new discoveries Bavink meant those aspects of modern physics according to which the microscopic or atomic level was ruled by chance alone. That chance meant for Bavink the absence of causality, and not merely our ignorance of causes, was suggested by his

---

[19] B. Bavink, *The Anatomy of Modern Science: An Introduction to the Scientific Philosophy of Today* (London: G. Bell & Sons, 1932), xiii + 683 pp. In the USA, it appeared under the title: *The Natural Sciences*.

[20] Bavink, *Science and God*, p. 132.

admission that a world steeped in the haphazard may seem much less in keeping with the traditional Christian view of the world as thoroughly ordered. Would a world of chance evoke, Bavink raised his typically German question, the recognition of the Creator in the same way in which the starry realm bespoke to Kant of a cosmic lawgiver? Bavink answered this question in the affirmative. His reason was that, after all, nothing happens or exists unless God directly brings it about. This meant, in Bavink's resolution of the theological question raised by Heisenberg's uncertainty principle, that the difference between classical physics and the physics of quantum mechanics was very simple. Within the former God created the original initial conditions in the Beginning; in the latter God keeps creating the initial conditions at every instant and for all events:

> In the literal sense, not a single quantum of action exists in the world which does not proceed directly and immediately from God. No natural law, not even a statistical one, compels its existence. Such a notion is just as meaningless as if we were to imagine that the statistics of railway accidents or marriages made one year compel those accidents or marriages taking place the next year, to occur. I think that the enormous liberation which this insight brings to religious thought makes it worth while to accept the apparent chance which it requires. For in truth, believers have always hitherto regarded chance as God's direct will (Matthew x. 29). This now becomes an evident fact for the chance in the

In the Twentieth Century

final elementary actions of existence is nothing other than the completely free decision by God.[21]

None of this should have surprised anyone who had carefully read the first line of the paragraph out of which this passage has been taken. There Bavink endorsed the "nominalist protests" against classical physics and against the inference that it was enough for the regular sequence of physical processes that their initial conditions be provided by God. That protest seemed to Bavink so well founded as to dispense with the need for going into "any great philosophical trouble of getting rid of objections to it."[22] It was, of course, another matter whether it was unreasonable to assume that God was capable of creating a physical realm with stable laws which He did not have to re-create at every moment but only had to conserve in its existence. This age-old Christian distinction between the erstwhile creation of things out of nothing and the conservation of the existence of things already created did not arise on Bavink's mental horizon. His claim that in the viewpoint endorsed by him chance was only apparent, rested on a theology harking back to Ockham who sought answer in miraculous interventions by God at every moment to essentially philosophical questions. That theology was eager in resorting to biblical phrases, such as the one (Matt 10:21) invoked by Bavink about sparrows none of which falls to the ground without the Heavenly Father's willing it. The fact that in many biblical passages the world is spoken of as firmly established and that even the endurance of God's covenant is

---

[21] Ibid., p. 136.
[22] Ibid.

asserted in terms of the endurance and unfailing regularity of his physical creation,[23] did not seem to have

[23] It is certainly not a coincidence that celebration of God's creation always precedes in the Psalms the celebration of the Covenant as, for instance in Psalms 18, 32, and 96. In Psalm 73, uncertainty about the Covenant owing to the destruction of Jerusalem is overcome with a portrayal of God's creative powers over the chaos. The same is true concerning the perplexity voiced in Psalm 88 about the eternity of David's throne. The unfailing outcome of salvation history is predicated on God's unfailing power evidenced in created nature in Isaiah 40:12-14, 21, 22, 28 and 40:24 and 45:12. Jeremiah's scoffing at idolatry with an eye on God's creative power (10:11-16) echoes the train of thought in Isaiah 40, but there is something very special in Jeremiah's references to the stability of nature in the context of his predictions of the Fall of Jerusalem. Instead of offering a "sign," that is, a miracle in the ordinary sense, as a proof that Yahweh's covenant with David will not fail, he points at the unfailing succession of day and night (Jer 33:19). Then Jeremiah repeats the same in reply to the dispirited slogan about the Lord's rejection of the two tribes (Jer 3:24-25). The impossibility of a created nature without stable laws is further emphasized in a similar assurance which Jeremiah offers following the destruction of Jerusalem (Jer 31:35-36). The crucial expression is "in spite of me," which is meant to convey the impossibility that Yahweh would ever abolish a law of nature he had originally set. Indeed the thinking of the prophets and psalmists shows no trace of a world being created anew at every moment. This is why the Bible, in spite of its assertion of the absolute sovereignty of God over nature, contains no hints about a difference between God's absolute and ordained power as understood by Ockham. Far removed is indeed the biblical world view from the one in the Kuran that provided powerful stimulus to Muslim orthodoxy which later inspired the occasionalism of al-Ashari and al-Ghazzali. Both were driven by the kind of "mysticism" that centuries later claimed as its victims Ockham and his countless followers, direct and indirect. For more details, concerning Bible and the history of science, see my publications, "The Universe and the Bible in

relevance for Bavink. The problems—scientific, philosophical, and theological—that transpire from the few pages Bavink devoted to miracles reappear in one way or another in all the subsequent discussions of miracles with an eye on modern physics. That the remaining pages will be mostly concerned with two books entirely devoted to miracles and modern physics is in part due to their wide availability to English readership. Another reason is that their respective authors are both professional physicists. One of them, William G. Pollard, is also an Anglican clergyman. The other, Donald M. MacKay, showed more than a passing interest in matters theological. Most importantly, their discussions are detailed and therefore provide their instructiveness in their own terms.

A chancy Providence

Pollard would have done better justice to his book had he called it not *Chance and Providence*[24] but "chance is providence" or, perhaps, "chance is your providence," though not "Providence is your chance." This is not to suggest that by Providence he did not mean most emphatically the one portrayed in the Bible. By chance he meant the randomness associated with quantum mechanics. He called it the "very task" or "primary thesis" of his book to show the full harmony of these two viewpoints.[25] It should therefore be no surprise that for Pollard

Modern Science," in *Ex Auditu* 3 (1987): 137-47; *The Savior of Science* (Washington, DC: Regnery-Gateway, 1988), pp. 56-66.

[24] W. G. Pollard, *Chance and Providence: God's Action in a World Governed by Scientific Law* (New York: Charles Scribner's Sons, 1958).

[25] Ibid., pp. 35 and 43.

quantum mechanics is the last word in physics. Conclusive for him had to be the failures, rather numerous by the mid-1950s, of efforts aimed at constructing a quantum mechanics with hidden variables, that is, a quantum mechanics which would rest on a mathematical formalism allowing for absolute precise measurements in principle at least. The conclusive character of quantum mechanics, which does not allow absolutely precise measurements of interactions, could be seen strengthened by experimental verifications, from the early 1980s on, of J. S. Bell's theorem of inequality.[26] Yet, however well-established may become the difference (inequality) between predictions made on the basis of classical statistics and quantum mechanical statistics, it would never dispose of the question whether an operational restriction on the precision of measurements is equivalent to an ontological incompleteness of the interactions to be measured. To anyone, such as Pollard, not facing up to this question, it is natural to state, as he does, that "the world is so *constituted* that the *ultimate* as well as present

---

[26] The real significance of Bell's theorem is that when explanations were sought for its experimental verification, the radically subjectivist character of the Copenhagen interpretation came once more into sharp focus. Prominent among those explanations was the view that instantaneous communication, a sort of "passion-at-a-distance," must be assumed to take place between photons traveling in opposite direction with the speed of light. But in that case, one must also assume instant communication among measuring instruments, and in terms of the Copenhagen philosophy of quantum mechanics, among observers. This situation is, however, equivalent to the utter futility of communication among all those who by adopting that philosophy have opted for solipsism in ultimate analysis.

In the Twentieth Century

characteristic mode of scientific *explanation* in all fields is statistical"[27] (Italics added).

Of the three words italicized (above) the first clearly carries an ontological meaning. Furthermore, only if that meaning is valid, and only then, that is, only if the world really embodies a basic randomness, is the use of the two other italicized words unobjectionable. To Pollard's credit, he is very conscious both of that logical connection and of the burden of proof it entails: "In order to establish my primary thesis that this is a necessary characteristic of scientific knowledge dictated by the nature of things rather than a merely temporary result of inadequate information, it is clearly necessary for us to probe much deeper than we have so far." Unfortunately, he does not fathom philosophical depths. In the same breath, and elsewhere too, he reasserts the fundamental ontological status of chance in the actual world "in which indeterminacy, alternative, and chance are *real* aspects of the fundamental nature of things, and not merely the consequence of our inadequate and provisional understanding."[28]

Ironically, this statement of Pollard is preceded by his dismissal of Einstein's disagreement with the celebration of chance on the basis of quantum mechanics. Pollard does so with the characterization of that disagreement as a mere "philosophical conviction." Philosophy fares indeed poorly in Pollard's book. Even elementary consistency is in short supply in connection with pivotal terms he uses. Thus he states about chance not only that "it cannot be the cause or reason for anything happening,"

---

[27] Pollard, *Chance and Providence*, p. 38.
[28] Ibid., pp. 43 and 54-55.

but also that "chance and probability in modern physics are . . . real and essential elements of the world which it describes."[29] The last statement implies, of course, the question of the value of scientific explanation. This crucial, philosophical problem is never met head-on by Pollard as if he had not heard of the countless books written on the subject both prior to and after the advent of quantum mechanics. Nor is the question, already aired here, about the legitimacy of inference from the operational to the ontological, so much as hinted at by him.

As one living in the physicists' world, he should not be too severely judged. The scientific community ignored countless warnings concerning that inference. If not the very first, certainly the most concise of those warnings was carried to the four corners of the scientific world through a letter that appeared in the December 29, 1930, issue of *Nature*, the leading scientific weekly. The concluding sentence of that letter, written by J. E. Turner of the University of Liverpool in connection with a prominent physicist's popularization of the chance world of atoms, contained more depth than much of the literature celebrating quantum mechanical chance: "Every argument that since some change cannot be 'determined' in the sense of 'ascertained,' it is therefore not 'determined' in the absolutely different sense of 'caused,' is a fallacy of equivocation."[30]

Whether Pollard perceived something of the sadly inadequate character of his reasoning on behalf of universal chance is a secondary matter. Nor is one to be appalled by the fact that as a scientist he fell completely

[29] Ibid., p. 104.
[30] *Nature* 126 (Dec. 27, 1930): 995.

under the sway of the extraordinary successes of quantum mechanical techniques and took them for basic and ultimate explanation. The same happened to countless colleagues of his, from the most outstanding to the most ordinary. What should seem to be especially instructive within the perspective of this essay is that he failed to perceive the devastating consequences devolving for Christian miracles from the very method he offered as their only safeguard. For underlying that method there seems to be an exaggerated measure of respect for science as it actually is, as respect that may undermine science as well as miracles (Providence) by the same stroke. Such undue measure of respect lurks between the lines, as Pollard states about his purpose of showing the full harmony of providence (Bible) and science (quantum mechanics). It is "to be accomplished in such a way that the essential integrity and unity of science, both as it is now and as in principle it may become, is fully preserved."[31]

Undoubtedly, a God who created human reason and is Reason himself deserves in full that *logike latreia* which Saint Paul enjoined (Romans 12:1) on Christians and Pollard may have had in mind. Such worship is incompatible with the slighting of anything that human reason can safely ascertain. By the same token that same kind of worship assumes as verity that there can be no contradiction between the historical revelation (be it in words or in deeds) of such a God and His self-revelation through nature which according to the same Saint Paul (Romans 1:20) is irrefutably clear, regardless of the resolve of some

---

[31] Pollard, *Chance and Providence*, p. 35.

to ignore it. But the non-existence of contradiction between revelation and reason can only be established if careful attention is given to the possible sources of misrepresentation of either or both. Contradictions are again bound to loom large if reason is limited to science, and even more so if the science of the day is taken for Science in its ultimate form. Neither science nor Revelation was served whenever God's basic way of action was taken to be equivalent to the workings attributed by *that* science to nature.

A fearsome boomerang

The story, several centuries old by now, is replayed with a new twist in Pollard's book. The great success of mechanistic or Newtonian science was a powerful motivation for casting God in the role of a clockmaker. But those theologians, whom Voltaire merely echoed in celebrating such a God,[32] were not eager to project him into the Bible. Pollard, however, is most emphatic in saying two things. One is that the idea of a God who suspends now and then the workings of the machinery of the world is "almost wholly unbiblical." The other is that only the notion of a God continually casting dice (that is doing what the chance of quantum mechanics is supposed to represent) is *wholly* and *alone* biblical. After

---

[32] While Voltaire's inference from the world as a clockwork to its Maker is well known, little attention has been paid to his far more expressive description of God as "the eternal machinist." Ironically, it is found in his half-serious, half-mocking *Traité de métaphysique* (1734). See *Oeuvres complètes de Voltaire*, ed. L. Moland (Paris: Garnier Frères, 1877-84), vol. 22, p. 223.

taking issue with those who speak disparagingly of "mere" chance, Pollard waxes dogmatic:

> To Einstein's famous question expressing his abhorrence of quantum mechanics, "Does God throw dice?" the Judeo-Christian answer is not, as so many have wrongly supposed, a denial, but a very positive affirmative. For only in a world in which the laws of nature govern events in accordance with the casting of dice can the Biblical view of a world whose history is responsive to God's will prevail.[33]

Before considering the allegedly biblical character of a dice-throwing God it should be worth considering the dice in question. Nothing would be more mistaken than to think of an ordinary die. The latter has six faces, eight corners, twelve edges, all definite parameters with such others as specific weight, elasticity, temperature, and so forth. Were God to be using such a die He would have to throw it but once. Its first and all subsequent bouncings off from a specific ground would strictly follow from the initial conditions of the first throw that could be known to God with complete accuracy. Nor would the case be any different were the various parameters of the die subject to statistical variations. What had already been said about statistical gas theory would apply here too. There one would still be within the framework of classical or "deterministic" physics. While we humans can only start from an average value of the parameters, to God all the individual cases of possible variations would be equally known and also their actual sequence as fully determined by the initial conditions.

---

[33] Pollard, *Chance and Providence*, p. 97.

Quite different would be the case of God throwing a die which is quantum mechanical in its Copenhagen interpretation. Such a die, radically different from the ordinary die, would display a random variability in the actually existing number of its parameters such as faces, edges, corners, etc. This has to be so long as one does not disavow the very core of the Copenhagen interpretation of quantum mechanics, that is, the logical somersault according to which an interaction that cannot be measured exactly cannot take place exactly or rather can take place only with an ontological defect in it. Instead of measurement one can, of course, refer to what it presupposes, the specification of parameters needed to carry it out. The necessary incompleteness of those specifications means, according to the Copenhagen philosophy, an ontological incompleteness.

To supply that defect, the Copenhagen camp or the overwhelming majority of physicists invokes chance or, as will be seen shortly, a short-sighted wizardry with mathematical operators. A theologian-physicist like Pollard, with equal allegiance to both of his professions, will, of course, invoke God too in addition to chance. The result is that all events in the physical realm (where all events are ultimately chance events according to the Copenhagen philosophy) become so many direct events actually performed by God, who alone can supply all parameters of the die which are (partly or entirely) unspecifiable by quantum mechanics and therefore (partly or entirely) non-existent according to that philosophy. If, however, such is the case, all natural events become miracles and all miracles become strictly natural events. Indeed, they become so many acts of creation out of nothing.

To his credit, Pollard minces no words: "It is an error to think of a miracle as being 'unnatural'." (According to him only the moral significance attributed by the faithful to very rare events turns them into miracles.)[34] To be sure, in another passage he restricts that sweeping statement to the "majority of biblical miracles." They "are the result of an extraordinary and extremely improbable combination of chance and accidents. They do not, on close analysis, involve, as is so frequently supposed, a violation of the law of nature."[35] He thinks that in such a way all miraculous healings listed in the New Testament are accounted for. As for large-scale nature-miracles, such as the one connected with the Exodus, they are still but natural coincidences for him. His exegesis is, of course, a rehash of ideas of liberal Protestant and modernist divines. He seems to follow them as he ascribes most biblical miracles to the hunger which "late elaborators" of those stories had for the miraculous.[36] In fact he retains only three events as miracles: the creation of all, the Incarnation, and Christ's Resurrection.

But is there a logical way of saving the reality of these three miracles while turning the Gospel account about many others into morality tales however exalted? One wonders whether Pollard thought of the heavy price paid by many liberal theologians for their being ashamed of miracles as so many violations of the "sacred" laws of

[34] Ibid., p. 117.
[35] Ibid., p. 83.
[36] Ibid., p. 115-16. Needless to say, Pollard does not name those late elaborators.

physics. Their fate is grippingly mirrored in the spiritual odyssey of Leo Tolstoy who took them for a guide. With his genius as a writer he could portray grippingly their starting point as well their state in the end. The former is succinctly given in the precept laid down in a notebook of his where the effort "to reinforce the teachings of Christ with miracles" is declared to be equivalent to "holding a lighted candle in front of the sun in order to see better." The end is illustrated by Tolstoy's harmony of the four gospels with so many passages cut out from the originals as to make a major biographer of his speak of it as "the Gospel according to St. Leo."[37] Better known is Tolstoy's novel, *Resurrection*, in which Christ's rising from the death is turned into a mere myth, shared, of course, communally.

Almost a hundred years later the Anglican bishop of Durham, Dr. Jenkins, served memorable evidence that the principle of demythologization inevitably turns, in the hands of a consistent devotee to it, the Gospel account about Christ's resurrection into a symbolic communal expression of hope in eternal life. Pollard's caveat that the Resurrection of Christ is an individual event and therefore cannot concern science, that is, quantum mechanics which deals with aggregates of events, is wholly beside the point. Christ's bodily resurrection does not come under the competence of quantum mechanics because it is a macroscopic event. Yet, in respect to the atoms constituting Christ's body, his resurrection would still be a chance event for which Pollard should have invoked God as the One who

[37] H. Troyat, *Tolstoy*, trans. from the French by N. Amphoux (1967; New York: Dell Publishing Co., 1969), pp. 476 and 483.

supplied ontologically, though "randomly," the parameters that could not be specified by quantum mechanics.

This is not the place to deal with Pollard's view that the ultimate truth of miracles rests with the faith of the community which endows a very rare event with a religious significance. Here let it suffice to say that such a falling back on communal faith would force the Christian to abandon the biblical injunction in terms of which he has to render a fully reasoned account of his faith and comportment. The same tactic also deprives the Christian of the possibility of challenging on a rational ground those who do not share his faith. They—agnostics, Jews, Muslims, Buddhists, Marxists, or Voodooists—are given by that tactic the same right to rest their case ultimately on their respective shared "faith." Clearly, what Pollard offers as a rational defense of Christian miracles is in fact a fearsome boomerang depriving that defense of all rationality.[38]

Pollard is but one of the many Protestants who fall back on that societal-fideist stance. In the same year (1958), M. Polanyi articulated the same stance on a vast scale in his *Personal Knowledge*. To his credit, Polanyi faced up to the specter of a boomerang as he quoted a statement of Lenin about the party spirit as the ultimate (enforced) foundation of Marxist materialism.[39] Whether the *free* sharing of faith (party spirit) in (Protestant) Christian society (Polanyi puts Communists and Roman Catholics in the same boat!) is a sufficient epistemological defense against that boomerang is a question that may have a reply in what Polanyi said about miracles:

[38] Ibid., p. 115.
[39] See Harper Torchbook edition, p. 245.

> Ever since the attack of philosophers like Bayle and Hume on the credibility of miracles, rationalists have urged that the acknowledgment of miracles must rest on the strength of factual evidence. But actually, the contrary is true: if the conversion of water into wine or the resurrection of the dead could be experimentally verified, this would strictly disprove their miraculous nature. Indeed, to the extent to which any event can be established in terms of natural science, it belongs to the natural order of things. However monstrous and surprising it may be, once it has been fully established as an observable fact, the event ceases to be regarded as supernatural.[40]

Polanyi's contention is faulty both historically and philosophically. Hume and others attacked miracles not on the basis of their factuality, but on the basis of the *necessarily* unchangeable character they attributed to the laws of nature and on the basis of the credibility of witnesses. Polanyi confuses the observation of a fact with its account in terms of natural science. The confusion arises from the fact that as a fideist philosopher and Christian, he could have no appreciation of the direct "commonsense" registering of facts as the simplest and plainest form of *knowledge*, and as the ultimate assurance about all knowledge, including its "tacit" kind.

As to Pollard, the disservice his explanation of miracles does to Christian faith derives ultimately from a distortion of the biblical portrayal of God's relation to the physical world created by Him. Of that relation Pollard mentions only God's full sovereignty over nature and His most intimate involvement in all its events. He

---

[40] Ibid, p. 284.

is silent about another no less noticeable aspect of that relation, namely, the constancy, endurance, stability, lawfulness, and consistency of the universe and its parts. As was already noted, the importance of that aspect is nowhere revealed more powerfully than in passages in which God's faithfulness to his covenant with David is supported with a reference to the faithfulness and stability with which the physical world created by Him follows its course and endures.

Unknowable coherence

That the biblical world view implies regularity and constancy was briefly recognized by Donald M. MacKay, author of the other book to be considered here.[41] He did so, however, in a way that runs the risk of destroying the natural knowability of those two cosmic qualities. The stability of the solid contents of the world is, according to MacKay, declared in the Bible

> to be a dynamic, contingent, stability. It is only in and through the continuing say-so of their and our Creator that they cohere or "hold together." He is the giver of being, moment by moment, to all the events in and through which we encounter the world of physical objects . . . But however uncompromisingly realist its tone, the Bible has no room for the idea of matter as something eternally self-sufficient or indestructible. The Psalmist may praise God for the stability of the earth and the reliability of the normal

---

[41] D. M. MacKay, *Science, Chance and Providence* (Oxford University Press, 1978). The book is the text of the Riddell Lectures delivered by MacKay at the University of Newcastle-upon-Tyne in March, 1977.

links between events on which our rational expectations are based, but the same Book of Psalms speaks of a time when the earth and the heavens will perish and will be changed "as a vesture" at the will of their Creator. In the end, for biblical theism, the only solid reality is God and what God holds in being.[42]

What this account leaves out of consideration is whether all those beings are kept by God in existence from moment to moment or are created anew out of nothing at every moment. MacKay's inveighing against what he calls the Greco-Medieval view of the cosmos consisting of "necessary natures" known by a priori reasoning is part of that partial vision.[43] It suggests that he cannot or is not willing to conceive of a "nature" which in itself is only one of the many possibilities available for God's creation but which by being a "nature" assures that it is maintained by God in existence in conformity to it. MacKay's oversight of the medieval philosophical and theological thinking (so critical of the Greeks on at least three crucial points) as the spark of the rise of modern science[44] is a corollary of his broader and distinctly Calvinist (Ockhamist) perspective.

This should be no surprise. From the moment Calvin (or Ockham) rose against apostolic succession,

---

[42] Ibid. p. 8.

[43] Ibid., p. 11.

[44] Ibid., p. 10. MacKay cites as authority M. B. Foster's article, "The Christian Doctrine of Creation and the Rise of Modern Natural Science," which was markedly dated when published in *Mind* in 1934, and woefully behind historical scholarship in 1977. The reason for this is Foster's ignorance of, or deliberate silence about, Pierre Duhem's gigantic and epoch-making presentation, in ten big volumes published between 1904 and 1916, of the medieval theological origins of Newtonian science. For details, see ch. 10, "The

they rose against ecclesial or sacramental continuity across time. To buttress philosophically their break with historic continuity they were instinctively pushed toward a world view steeped in discontinuities, that is, toward a Nature without "natures." No wonder MacKay is pushed by the logic of his own position to asserting that continuity and objectivity in nature can only be known by one's surrendering to biblical revelation, or rather to MacKay's reading of it. Thus, in discussing the lure of quantum theory which challenges the distinction between the observer and the observed, and the lure of relativity theory which challenges the validity of a single description of the world valid for all observers, MacKay concludes: "True our Christian scientist in God's world may have no access to the Creator's eye view of his situation; but because he knows that he is under judgment by that criterion, he is saved from the trap of confusing relativity with a denial of objectivity."[45]

It is the same condition of standing under the judgment of biblical revelation as he interprets it that saves MacKay's believing scientist from the pitfalls of quantum theory. This declaration of MacKay is all the more instructive because he does not invoke revelation or biblical perspective as an indispensable ground for recognizing that Einstein's opponents "had no rational grounds for claiming that the absence of an *observable*

Historian," in my *Uneasy Genius: The Life and Work of Pierre Duhem* (Dordrecht, London, Boston. Martinus Nijhoff, 1984; paperback reissue, 1987). On Foster, see also my essay, "Telltale Remarks and a Tale Untold," in *Creation, Nature and Political Order in the Philosophy of Michael Foster (1903 - 1959)*, ed. C. Wybrow (Lewiston, NY: The Edwin Mellen Press, 1992), pp. 269-96.

[45] MacKay, *Science, Chance and Providence*, p. 19.

causal precedent for an event meant that it had *no* causal precedent.'[46] But no sooner had MacKay seemed to sight the ontological perspective of the problem, he lost sight of it, or perhaps he deliberately tried to cover it up with a theological smokescreen which is, however, rather transparent. For the question is not whether God can play dice in a manner worthy of Him. The question is once more about the way in which the parameters or details that cannot be specified by quantum mechanics become presently existing in order to let things go. MacKay tries to resolve that question on the basis that God is most directly involved in all events as a Sovereign Lord over all creation. But in doing so he merely sidesteps the problem which once more surfaces as he states: "From a biblical standpoint it would be equally mistaken to argue that if there were no causal precursor for an event, then its Creator must be thought of as 'playing dice'."[47] For the absence of a causal precursor, or even the partial absence of such a precursor, is an ontological gap. Does it fall upon MacKay's "biblical God" to fill that gap with continual instantaneous creations? That such is indeed the thrust of MacKay's reasoning may be surmised from his further talking around the problem without being ready to meet it head on:

> The God of biblical theism is beholden to none to account for his creative agency. If he freely wills into being a succession of events in which one half of the sub-microscopic details at any time are unspecified by their precursors, this would involve no inconsistency with his character, still less with his sovereignty, as portrayed in the Bible. Belief in a sover-

[46] Ibid., p. 30.
[47] Ibid.

In the Twentieth Century

eign God does not in the least entail a belief that there *must* be "hidden physical variables" sufficient to determine the behaviour of electrons on the basis of precedent. For biblical theism all events, equally, with or without precursors according to precedent, need God's say-so in order that they occur at all. The choice of "God or Chance" is simply not a meaningful alternative, if "Chance" is meant in the scientific sense. As the Book of Proverbs (ch. 16, v. 33) has it: "The lots may be cast into the lap, but the issue depends wholly on the Lord."[48]

Whatever the appropriateness of that particular biblical passage, MacKay's banking on God's sovereignty seems to be very inappropriate. The issue is not that God is sovereign but whether He is at least as rational as human beings are. The latter can clearly recognize the difference between being and non-being. It is that difference which is the real issue concerning the dispute about the theory of hidden variables. Whether most professional supporters and opponents of that theory have realized this is irrelevant. As was already stated, physicists have for centuries been apt to limit their vistas to quantitatively exact measurements and this is why they have taken it for mechanical (and implicitly for ontological) causality. Insofar as hidden variables (or any parameter of mathematical physics, for that matter) need an ontological substratum, the biblical God, whose self-revealed name I AM WHO AM is ontology incarnate,[49] is

---

[48] Ibid.

[49] This is not to suggest, to recall a memorable remark of E. Gilson, "that the text of Exodus is a revealed metaphysical definition of God: but if there is no metaphysics *in* Exodus, there is nevertheless a metaphysics *of* Exodus," vastly articulated by the Fathers of the Church and medieval philosopher-theologians. *The*

bound to provide it in one way or another. MacKay seems to suggest that He does so by continually creating out of nothing at least one half of the ontological specifications of all atoms to fill the ontological gaps created by quantum mechanics or rather by its Copenhagen pseudo-philosophy.

In a manner characteristic of the entire Ockhamist tradition that heavily conditioned the Reformers' theologizing (for all their dislike for philosophy), MacKay goes straight to God's sovereignty as if this could dispose of a plain question whether the existence or non-existence of certain things is meant by some staple phrases of quantum mechanics. No wonder that he sees but the veneer of the fallacy of setting up Chance as a kind of anti-deity standing for chaos and absence of meaning.[50] He fails to see that the basic issue about chance is whether it can be a substitute for reality, be it the reality of physical parameters that "do not exist" because quantum mechanics cannot specify them. Understandably, MacKay did his utmost to make it appear that his train of thought did not force him to charge God with the task of supplying the reality of those unspecified parameters with special creative acts performed every split second. His fellow scientists would have been taken aback, though not for the right reason. The latter, not a matter of theology but of plain philosophy, could not,

---

Spirit of Mediaeval Philosophy, trans. A. H. C. Downes (New York: Charles Scribner's Sons, 1936), pp. 51 and 433-434.

[50] One wonders whether Emerson, who scorned "the ancients, [who] struck with this irreducibleness of the elements of human life to calculation, exalted Chance into a Divinity" (*Emerson's Complete Works* [Boston: Houghton, Mifflin, 1888], vol. 2, p. 71) would not now choose for target some moderns who are theologians to boot.

however, be seen by the ones who had bartered sound reasoning about reality for the hollow glitter of mathematical operators which give no certainty about anything real, let alone about miracles and Providence in human history.

CHAPTER THREE

Perennial Perspectives

The only starting point
Certainty about real events or things, usual or unusual, can never begin with science, and not even with "the Lord's quantum mechanics"[1] as Schrödinger once spoke

---

[1] E. Schrödinger, *What is life? and Other Scientific Essays* (Garden City, N.Y.: Doubleday, 1956), p. 83. It is only fair to note that a generation later no less an erstwhile architect of quantum mechanics than Dirac conjured up the vision of its future form which "will have determinism in the way that Einstein wanted" though at present, physicists may still have to endorse its indeterminateness à la Bohr "especially if they have examinations in front of them." This startling "confession" of Dirac, made at the Jerusalem Einstein Centennial Conference in 1979 and duly reported by one of the participants, R. Resnick, in *Journal of Chemical Education* 52 (1980): 860, may help one see a prophetic ring in the words of H. Landé, another architect of quantum theory, and one of the few cries in the philosophical wilderness produced by Bohr and his vast coterie: "Using the age-old skepticism of philosophers as to the reality of the external world to serve as a cover for our temporary ignorance and indecision, is the policy of 'if you can't explain it, call it a principle, then look down on those who still search for an explanation as unenlightened." *From Dualism to Unity in Quantum Mechanics* (Cambridge: University Press, 1960), p. 56. This is not to suggest that Landé really saw the difference between exact measurements and ontological causality in the things of the external world as objects of immediate knowledge. But at least he did not turn against his right cognitive instincts. For a very recent example of

in quasi-mystical awe of his own specialty. Science rather presupposes real things in order to ascertain their quantitative properties. Science cannot provide any of those things, and not even those properties of theirs. Of course, nothing reveals so forcefully the reality of things as their limitedness which has many quantitative components. But whether they, or the things in which they are embedded, do exist or not is not a scientific question. Being a question about the real, it cannot be answered except by a philosophy which provides the perennially proper place for the question.

To be sure, even idealist philosophies make claim to reality, to say nothing of the philosophies known as rationalism, empiricism, sensationism, and pragmatism. They all claim to be *the* place for the real. But not the proper or primary place. They are indeed betrayed by their labels, which are almost always the choice of their chief articulators. Those labels invariably relate to an aspect of the real insofar as it can be conceptualized, sensed, tested, manipulated and so forth, but never to the very reality of things.

None of those philosophies would carry their special labels if the very start of their program and method were an unconditional acknowledgment of external reality. Only that acknowledgment is a guarantee of knowing that reality with certainty. This may appear an arguing in a circle. Actually, it is the only starting point which can save one from arguing ever in such a way. It is a consciously and methodically-taken starting point. A

the same ambivalence, see F. Rohrlich, "Reality and Quantum Mechanics," *Annals of New York Academy of Sciences* 48 (1986), pp. 373-81.

chief recommendation of that method is that all other philosophical approaches to knowing the real with certainty have turned out to be so many seeds of doubt about reality or means whereby the thinking man found himself cut off from the external world. Cartesian rationalism that aimed at complete certainty (equating it with mathematics which is a series of tautologies, however useful) led to Spinoza's perplexity about any and all finite real things.[2] Francis Bacon's empiricist "instauration" of a new age in thinking revealed in Hume's hands the fragmentation through it of all judgment about reality. Kant's effort to restore certitude through the a priori character of mental categories led in Fichte's hands to the exaltation of the will and, in Hegel's hands, to the divinization of the individual ego, the least reliable commodity one can think of. The sensationism advocated by Mach locked him in solipsism which is undoubtedly the highest conceivable measure of certainty although not communicable. As to pragmatism, its chief spokesmen, William James and John Dewey, would in vain try today to disavow the uncertainty which it has generated about everything except, of course, one's selfish and all too often very transient success.

The other chief recommendation for taking the certainty of knowing external reality for the starting point in all philosophy is that any refutation of it implies knowledge of that type. Thus to argue that a specific registering of a fact, thing, or event was a mere hallucination, one must assume that it is possible not to be under its influence in registering this or that fact. The

---

[2] See note 30 to Ch. 1.

same holds true about the argument based on any partial deception or error of one's senses or on any exaggerated claim about the extent of one's observations. Those "critical" philosophers who have succeeded in spreading the belief that nothing can be known unless first critically proven have in fact assumed this "proven" knowledge without first critically proving it. Moreover, just as colors cannot be discoursed upon in terms of non-colors, the knowledge of external reality cannot be proven in terms of knowing only one's mind, "critical" or not.[3]

The certainty of facts

This is basically *all* that is needed to show the certainty of facts called miracles. The *all* in question may sound very little, but actually it is co-extensive with that largest entity called the world of the real, and also co-extensive with all reasoned discourse relating to it. In a sense, that *all* is very restricted as it is ultimately reduced to the evidence of one's unaided senses. This may appear ridiculously little in an age of science that probes such realms of the very small and the very large that are inconceivably beyond the reach of the senses. It should not, however, be forgotten that the ultimate certainty of all the esoteric findings of science in the farthest reaches of space and in the deepest layers of matter rests on the reliability of the senses that register the position of ordinary pointer needles. This is what no less an "idealist" physicist than Eddington recognized when he stated that "molar physics has the last word in observation for

---

[3] This is in substance Gilson's objection to the very nerve center of the Kantian theory of knowledge according to which knowledge begins with its own criticism.

the observer is molar."⁴ And if the physicist takes no stock of this, he can embroil himself in the kind of embarrassment which left speechless for a moment the famed astronomer-cosmologist W. H. McCrea. After being heard to state in a lecture that the star images seen through the telescopes have a strict relation to reality only insofar as they are sensations on the retina, he was asked in the question-answer period: "Would you also hold the same about the reality of the wall which you are facing?" His answer, "I am not really sure," speaks for itself.⁵ Not even that much comment is deserved by the inconsistency of those astronomers (some world-famous) who after boasting of their solipsism during dinner,⁶ do not blush as they spend the night looking through their telescopes.

Immediate direct observation of things and the certainty of that observation (or at least the certainty with which it can be corrected or improved) is the rock bottom basis not only of philosophy but of science as well. In view of this and of what already has been said about the true status of scientific laws, it should not be difficult to perceive the disingenuousness of the indignation with which miracles are denounced as violations of

---

[4] A. S. Eddington, *The Philosophy of Physical Science* (New York: Macmillan Company, 1939), p. 77.

[5] The objection was made by me as a participant at the Second International Colloquium held at the University of Denver in November, 1974, whose proceedings are available in *Cosmology, History and Theology*, eds. W. Yourgrau and A. D. Breck (New York: Plenum, 1977).

[6] I have in mind A. Sandage, who was sitting across from me at the dinner given in his honor prior to his lecture at my university, Seton Hall, South Orange, N.J., in the spring of 1971.

the laws of nature. The indignation is essentially a clever form of the strategy: attack is the best defense. But if it is impossible to start a march (physical or mental) with the second step, concern about the laws of nature should give second place to concern about man's ability to register things and events with certainty. And since without that ability nothing can be known about the laws of nature, the chief intellectual concern should be not so much about the possible violations of the laws of nature as about the actual violation, if not plain rape, of man's mind whose natural function is to know reality with immediate certitude.

Such a rape is committed when individuals reporting extraordinary events, and in fact lay down their lives on behalf of their witness, are declared at the outset to be hotheaded enthusiasts, uncritical minds, or plain fakers. This is done on the patently dogmatic ground that nature cannot change its course. Those taking that ground rape their own intellect to the point of declaring that they cannot even have one. A startling admission of this came from such a prominent spokesman of the absolute unchangeability of "nature's laws" (a form of sheer materialism) as J. B. S. Haldane:

> If my mental processes are determined wholly by the motions of atoms in my brain, I have no reason to suppose that my beliefs are true. They may be sound chemically, but that does not make them sound logically. And hence I have no reason for supposing my brain to be composed of atoms.[7]

---

[7] J. B. S. Haldane, *Possible Worlds and Other Essays* (London: Chatto & Windus, 1927), p. 209. This passage is praised by C. S. Lewis in his *Miracles: A Preliminary Study* (5th ed.; New York: The Macmillan

The defense of miracles done with an eye on physics should include a passing reference to meteorites. Characteristic of the stubborn resistance of scientific academies to those strange bits of matter was Laplace's shouting, "We've had enough such myths," when Pictet, a fellow academician, urged a reconsideration of the evidence provided by "lay-people" as plain eyewitnesses.[8] Laymen were they in the sense that they had no telescopes, no training in celestial mechanics, no knowledge of trajectories, azimuth, right and left ascension. But they could register with absolute certainty that a fiery body had just hit the ground nearby and could unerringly distinguish its still warm stony remains as something not

---

Co., 1955, pp. 28-29) as "the shortest and simplest form" of the argument that "the Naturalist cannot condemn other people's thoughts because they have irrational causes and continues to believe his own which have (if Naturalism is true) equally irrational causes." Yet Lewis does not quote the crucial second sentence in the passage. The same is true of the use of that passage in *Miracles in the Critical Mind* (see note 6 to Ch. 2) by C. Brown, who obviously took (p. 230) the passage "on faith" from Lewis and did not care to look up the original.

[8] During the decades of the Enlightenment, the resistance of those Academies, influenced largely by the Académie des Sciences in Paris, became so great as to result in the discarding of all meteorites from museums. Fortunately, the bishop of Zagreb refused to be influenced by that "scientific stampede" when large meteorites fell in nearby Hraschina in 1751. He ordered his consistory to collect from eyewitnesses sworn statements, which he sent, together with the meteorites, to the Emperor in Vienna. The bishop "did the most reasonable thing that," in the words of Chladni, the founder of the science of meteroritics, "could be done under the circumstances." See E. F. Chladni, *Ueber Feuer-Meteore und ueber die mit denselben herabgefallenen Massen* (Vienna: J. G. Heubner, 1819), p. 245.

belonging to the soil around it. That such a kind of witnessing stands in its own right was the point recognized by a doctor on being confronted with the objection of a colleague who insisted that the wide-open fracture below the left knee of Pieter De Rudder (1822-1898), the subject of possibly the most startling cure related to Lourdes, could not be accepted for a fact because the two ends of the broken bone protruding through the skin had not been certified by a medical commission. The reply of that rightly indignant physician, "it does not take a tailor to see that a coat is full of holes,"[9] contains an instructiveness that is practically inexhaustible.

The case of miracles

The case of sighting meteorites, however extraordinary, is not the same as the case of miracles. Unlike meteorites that repeat themselves, any given miracle is a strictly individual event that cannot be expected to occur again. Its verification, even in the case of a fresh miracle, is essentially one involving the historical method with its reliance on direct witnesses, on indirect observation, and circumstantial evidence. This is why ancient miracles as objects of historical verification are a much more difficult matter than are recent miracles. Only upon the latter does beat "the bright light of modern history," to recall a felicitous expression of that famed Jewish novelist Franz Werfel in his introduction to *The Song of Bernadette*, his moving narrative of the first crucial three months at Lourdes.[10]

---

[9] Quoted in L. Monden, *Signs and Wonders*, p. 244.
[10] F. Werfel, *The Song of Bernadette*, trans. L. Lewisohn (New York: The Viking Press, 1942), p. 7. Werfel adds: "And their truth has

For all his certainty about the miracles of Lourdes, and for all his gratitude to the Lady of its Grotto and to the memory of her humble maidservant Bernadette Soubirous (whose body he knew to lie incorrupt in a glass casket in Nevers), Werfel did not become a Christian, a Catholic. As to Alexis Carrel, who received in 1912 the Nobel Prize for his study of the rate at which wounds heal, he first went in 1903 to Lourdes,[11] where incredibly fast healing of festering wounds had by then been attested for almost half a century. Yet, it was not until 1940 or so that Carrel was able to get rid of all his agnostic reservations and become a Christian, a Catholic, although long before that he had known of the powerful argument that reason could forge from an attentive consideration of those cures. The argument had already been voiced on more than one occasion when in 1909 Teilhard de Chardin cast it into a classic form with his powerful prose:

> If a common antecedent for the cures could only be discovered; if we could extract from all these au-

been confirmed by friend and foe and by cool observers through faithful testimonies."

[11] Carrel's account of that trip, written in the third person, was published five years after his death under the title, *Voyage de Lourdes: suivi de Fragments, de Journal et de Méditations* (Paris: Plon, 1949). The introduction provides incontrovertible evidence of Carrel's profession of Catholic faith and of his reception, in full mental strength, of the last sacraments of the Church. It is in that light that one should see Carrel's references to Lourdes in his widely available *Man the Unknown* (1935; New York: McFadden 1961), p. 101. A discussion of Carrel's relation to Lourdes, based on the archival evidence in the Medical Bureau there, is available in my introduction to the re-edition of his *Voyage to Lourdes* (Royal Oak, MI: Real View Books, 1994).

thentic facts something which marks them off or conditions them! But we find only this: *Lourdes*; and it is not the Lourdes imagined or hoped for in the excitement of pilgrimages . . . but it is Lourdes alone—Lourdes, a naked and objective reality, to which is attached a mysterious virtue, independent of anything the sick and the praying crowds can take there. If the cures of Lourdes were characterized by any family likeness, attached to one category of diseases or appeared under determinate circumstances of time or place, I might invoke with show of reason, some magnetism, some appropriate vibration with which the human body would enter into a vivifying resonance. The precise cause would escape me, but a certain regularity in the phenomena would assure me of the existence of this cause and entitle me to imagine it. But there is nothing of the kind . . . effects follow each other without apparent rule. These cures are distributed as if by chance, and sometimes there are alarming relapses. In all truth, what renders Lourdes altogether extra-medical is less what occurs there than the manner in which the prodigies take place. If what happens there astonishes the scientists, the way it happens is absolutely beyond him.[12]

---

[12] P. Teilhard de Chardin, "Les miracles de Lourdes et les enquêtes canoniques," *Etudes* 118 (1909), pp. 161-183; for the passages quoted, see pp. 176-177. They follow Teilhard de Chardin's consideration of the possibility whether the cause of those miraculous cures could be attributed to a physical force still unknown. Father Teilhard is not quoted as an *authority*. No book or article, by an author however prominent or popular, deserves such status. But the author of every book or article deserves protection against being misrepresented, a point also appropriate à propos the events and facts connected with Lourdes. C. Brown, the author of

One of the purposes of this essay was to call attention to the role which the recognition of "naked and objective reality" (of plain facts) plays in the philosophy that alone can do justice to facts be they so extraordinary as to be called miracles. The chief recommendation of that philosophy is that it alone can cope also with the facts of ordinary life as well as with the facts which science carefully isolates for its purposes. For even in the systematic isolation or carefully controlled conditions which science demands for its facts, their usefulness ultimately depends on the reliability of plain human witness about them. Without that witness not only the vast enterprise known as scientific endeavor would lose its claim to truth, but also the far more vast social life would be deprived of its right to justice.

Courts of all levels, governments of all jurisdiction, depend on witnesses and their plain witnessing[13] and so

---

Miracles and the Critical Mind (see note 6 to Ch. 2) can be taken to task in both respects. He presents (p. 349, note 18) L. Sabourin's work, *The Divine Miracles Discussed and Defended* (Roma: Officium Libri Catholici, 1977) as one drawing heavily on D. J. West's *Eleven Lourdes Miracles* (London: Duckworth, 1957) and adds that West "complains of the lack of thoroughness in the investigation of cases, which he believes does not preclude wrong diagnosis and natural remission of the illnesses." The fact is that Sabourin called attention to the fact (p. 158), not mentioned by Brown, that West did not meet any of those cured, nor did he study all the documents relating to them. Further, Sabourin pointedly noted that as one "especially interested in psychical research, West will more or less consciously be inclined to find what he was looking for: psychical explanations of exceptional cures."

[13] Last but not least, Congressional hearings, that have been making headlines ever since Watergate, derive their value from the trust placed in the testimony of witnesses.

do laboratories. In none of those forums can a discrimination against plain witnessing of unusual facts be condoned or else the most important cases may be prejudged and the only avenues for progress be blocked. Had Oersted refused to believe his eyes when they noted that the magnetic needle which he placed under a live wire turned in a direction which he believed to be impossible, the discoveries of Faraday and Maxwell might not have followed as they did. The discovery of the world of atoms depended on Roentgen's chance witnessing the formation, that was not expected to happen, of the negative image of a key on a photographic plate. Far more importantly, would Newtonian science have developed at all if Kepler had not unconditionally trusted Tycho Brahe's eyes in making countless naked-eye observations about the positions of the planet Mars?

Luckily for science, scientists relatively rarely brush aside reports about a really *new* case with the remark that it cannot be different from a thousand or so cases already investigated. But whenever a scientist (the psychoanalyst A. Adler is a case in point), makes such a remark, one is bound to react as did K. R. Popper: "And I could not help myself saying: 'And with this new case, I suppose, your experience has become thousand-and-one fold'."[14] This reaction, sarcastic but well deserved, which is not readily recalled in the corridors of psychoanalysis, is precisely the rejoinder that is to be offered whenever facts are dismissed because they evoke the specter of miracles.

---

[14] K. R. Popper, *Conjectures and Refutations: The Growth of Scientific Knowledge* (London: Routledge and Kegan Paul, 1962), p. 35.

The witnessing of facts is, of course, to be coupled with a willingness to face up to the consequences of the fact witnessed. If the author of the Book of Joshua meant literally the stopping of the sun "in the middle of the sky" and staying there "for a whole day" (Joshua 10:13), then one would have on hand astronomical consequences that even from a distance of three thousand years could be verified. No other biblical miracle would pose a similar problem.[15] Such physical miracles as the multiplication of the bread, the changing of water into wine, Christ's and Peter's walking on the water, represent disturbances that cannot be detected from a distance of two thousand years. This would be the case even if they were to be contemporary events. The reason for this is not so much the relative minuteness of the physical effect they represent, but the impossibility of making the scientific apparatus ready for the event. This is not to say that there would not be countless men of science ready to stand by with all sorts of sensors to register the physical parameters of a physical miracle, including the rapid healing of festering wounds, of broken bones, of collapsed lungs, and of lumps of cancerous tissues. But those men of science would wait in vain for an invitation from on High or from any of the Almighty's saintly agents. Miracles are not for order. They never were.

The demythologization of science

This is the only point about miracles which puts the believer at a disadvantage. Humiliating it may be, but a

---

[15] For a discussion of Joshua's miracle and other major physical miracles in the Bible, see my *Bible and Science* (Front Royal, VA: Christendom Press, 1996).

humiliation fully consistent with the humble framework in which the two Covenants were offered to man across the span of almost two millennia. An insignificant corner of the earth was chosen to be the scene of both Covenants. The consent recipient was a people that should seem most insignificant compared with the cultural, artistic, and organizational magnificence of great neighboring civilizations. There is, of course, inside that humiliation a silver lining which is nothing short of a miracle: a unique interpretation of history, human and cosmic, physical and moral, compared with which all other interpretations, ancient or recent, are a poor second. That ultimately the rise of science was sparked by that interpretation[16] is hardly a point to let drift from focus in the often theatrical confrontation of miracles with science.

No humiliation is involved in the fact that miracles are never automatically overwhelming proofs. They represent the challenge of external reality, not of axioms of logic.[17] That true miracles are never coercive, whatever

---

[16] The essence of that interpretation is the linear notion of cosmic and human history, with a most specific absolute beginning toward a no less specific end. It was the keeping in focus of that absolute beginning that enabled Buridan and Oresme to formulate in the fourteenth century the idea of inertial motion and impetus (momentum). This great discovery, which Duhem made known around 1907, is discussed in my works *Science and Creation: From Eternal Cycles to an Oscillating Universe* (1974; 2nd rev. ed., Edinburgh: Scottish Academic Press, 1986), pp. 231-241; *Uneasy Genius: The Life and Work of Pierre Duhem*, pp. 390-400 and 428-429; and *The Savior of Science*, pp. 50-59.

[17] This should seem an all-important point in an age which has increasingly equated proofs with mathematical demonstrations. There is no demonstration or formula of logic that would cope with

their occasional impact on skeptics and scoffers, is their chief recommendation. A dispensation would never be truly divine that would take man's freedom away because such a dispensation would not also be fully human. Clearly, it all depends on the perspective or, to use the technical term, philosophy or epistemology. That all knowledge, not only the knowability of miracles but everything else, depends on that perspective, is implied in the recognition that, to borrow a forceful phrase from a famed analysis of the origins of modern science, "The only way to avoid becoming a metaphysician is to say nothing."[18] What the author actually meant was the very opposite of the meaning which is usually ascribed to metaphysics: the art of bartering facts for ideas. Unfortunately, the author in question did not know of the only metaphysics, Aristotelian-Thomistic metaphysics, that begins with the recognition of facts and claims in fact that all the rules (categories) of man's mental operations are a distillation from his registering of facts.[19]

To approach any subject, be it the subject of miracles, in any other way will land the mind in mirages as witnessed by the despair of modern man about his

---

the relation between the knower and the external objective world known by him. Therein lies the ultimate futility of reductionism based on science.

[18] E. A. Burtt, *The Metaphysical Foundations of Modern Physician Science* (1924; rev. ed., Garden City, NY: Doubleday, n.d.), p. 227.

[19] The point is made most explicitly by Aquinas: "Some believed that the *intellectus agens* to be nothing else but a habit in us of undemonstrable principles. But this cannot be so because we know even the undemonstrable principles by abstracting [them] from the sensory [evidence]." *Quaestiones disputatae de anima*, I, art. 5, ad Resp. (Turin: Marietti, 1953).

intellect. That man will find help only from those Christians who have not lost sight, even for a moment, of the truly realistic epistemology. Such Christians and only such can fully seize their intellectual opportunity which is offered by those unbelievers who at least admit the fact of certain extraordinary events, though not their miraculous character. They are at one with T. H. Huxley who urged that unreserved attention be given to all facts, however extraordinary. They would emulate also that Huxley who, following the death of his first son at the age of seven, firmly declined the comfort of Christian perspectives. Huxley did so with a profession of faith in the facts of nature as seen by science as he understood it:

> Science seems to me to teach in the highest and strongest manner the great truth which is embodied in the Christian conception of entire surrender to the will of God. Sit down before fact as little child, prepared to give up every preconceived notion, follow humbly wherever and to whatever abysses nature leads, or you shall learn nothing. I have only begun to learn content and peace of mind since I have resolved at all risks to do this.[20]

The defender of Christian miracles should, of course, be able to demythologize that notion of science which Huxley in the same context made an object of worship with his eyes fixed on the inverse square law, in obvious ignorance of its not entirely "scientific" provenance:

> It is no use to talk to me of analogies and probabilities. I know what I mean when I say I believe in the

[20] L. Huxley, *The Life and Letters of Thomas Henry Huxley* (London: Macmillan, 1900), vol. 1, p. 219.

law of the inverse squares, and I will not rest my life and my hopes upon weaker convictions. I dare not if I would.[21]

But even with that demythologization of science done, the Christian defender of miracles must tirelessly return to them insofar as they are facts, and insist that they be faced with the openness of a child. He can do no more than that teenage peasant girl, Bernadette Soubirous, whose mental aplomb under endless questioning was no less a miracle than the cures her visions had triggered. To a visitor pressing her with doubts about those visions, she gave this reply of astonishing balance between the certainty of facts and the difficulty of convincing others about them: "Je suis chargée de vous le dire, je ne suis pas chargée de vous le faire croire."[22]

Facts and convincing

This is all a Christian can do about miracles. He has to reassert them as facts in all their details and context but he should under no circumstances confuse the firm presentation of facts with the art of convincing others about them. About miracles, however factual, conviction

---

[21] Ibid., pp. 217-218.

[22] "It is my duty to tell it to you, it is not my duty to make you believe it." Quoted in A. Olivieri, M.D., and Dom Bernard Billet, *Y a-t-il encore des miracles à Lourdes? 21 dossiers de guérisons* (new ed.; Paris: P. Lethielleux, 1979), p. 58. She gave a no less astonishing glimpse of her utter certitude about her sense perception when challenged by the abbé Corbin's searching question: "What would you have replied if the bishop of Tarbes had judged that you were mistaken?" She replied: "I would never have been able to say that I did not see and did not hear." Quoted in R. Laurentin, *Vie de Bernadette* (Paris: Desclée de Brouwer, 1978), p. 124.

is a matter of God's grace which, however, has an intimate tie to facts, however miraculous, that can be heard, seen, and touched. It is these very terms that are the object of a perception which is as sensory as it is also an understanding or *episteme*. A biblical proof of this is on hand in the very beginning of the first epistle of John. There the entire Christian message is cast into a realist epistemological frame:

> This is what we proclaim to you:
> what was from the beginning,
> what we have *heard*,
> what we have *looked* upon,
> and our hands have *touched*—
> we speak of the word of life[23] (Italics added).

This kind of epistemology stands somewhere in the middle between the classic extremes of positivism and idealism. In positivism, the tangible facts can never lead to metaphysical heights, let alone to heights where the Word of Life is heard. In idealism, the metaphysical heights are not supposed to be rooted in that reality which human touch and sight alone give access to. Only when imbued with a median epistemology will Christians be liberated from a veneer of sophistication about miracles which is but a throwback to a leery Humean skepticism. Only then will they instinctively avoid either ending or beginning their discussion of miracles with the despondent sigh, a transparent admission of an intellectual failure of the nerve: "Miracle was once the foundation of all apologetics, then it became an apologetic

---

[23] This statement of John, offered in part as a shield against the Hegelians (Gnostics) of his time, is of course a perfect echo of the thoroughly realist tone of the entire biblical revelation.

crutch, and today it is not infrequently regarded as a cross for apologetics to bear."[24] Only when Christians relearn to glory in their minds as an organ whose natural function is to have certainty about facts and things will they be able to derive intellectual glory from miracles. On that certainty and on it alone can that intellectual platform be built which provides proper perspective about science, about miracles, and even about God insofar as He can be grasped by that reason which makes man a being created in His very image.

---

[24] R. Seeberg, "Wunder," *Realencyklopädie für protestantische Theologie und Kirche* (Leipzig, 1908), vol. XXI, p. 562. Immediately preceding this statement of Seeberg is his disapproval of Christian theologians who accept Rousseau's remark that the truth of miracles conditions the truth of Christian revelation. C. Brown, who uses Seeberg's statement as the motto of the first chapter of his *That You May Believe: Miracles and Faith Then and Now* (Grand Rapids, MI: Wm. B. Eerdmans, 1985) concludes on a not much different note by quoting the words, "My grace is sufficient," addressed to Saint Paul looking for a cure of a still unidentified disability of his. Brown should have added that for the same Paul miracles were an essential help in his spreading the message about God's grace. Suffice it to recall here his healing of the boy who fell from the upstairs window, his immunity to the viper's bite on the island of Malta, and the escape of all 276 on board the ship that fell apart nearby, after being tossed around by a violent storm for a full fortnight.

Index of Names

Adams, J., 38
Adler, A., 88
Alexander, S., 33
Ampère, A. M., 13
Aquinas, Thomas, Saint, 91
Aristotle, 28
Ashari, -al, 29, 56

Bacon, F., 23, 38, 79
Bavink, B., 51-55, 57
Bayle, P., 4, 68
Bell, J. S., 58
Benedict XIV, Pope, 5
Bergson, H., 33
Billet, B., 93
Bloch, E., 33
Bohr, N., 48, 77
Born, M., 48
Boyle, R., 33
Brahe, T., 88
Brown, C., 44, 83, 86, 95
Buffier, C., 21
Bultmann, R., 7, 11-13, 33, 45
Bunyan, J., 19

Buridan, J., 90
Burtt, E. A., 91

Calvin, J., 70
Carrel, A., 85
Chesterton, G. K., vii, 30
Chladni, E. F., 83
Compton, A. H., 35-36
Corbin, F., 93

De Rudder, P., 84
Descartes, R., 3-4, 27-28, 38-39,
Dewey, J., 79
Dibelius, M., 11-12
Dirac, P.A. M., 77
Driscoll, J. T., 51
Duhem, P., 70, 90

Eddington, A. S., 50, 80
Edison, T. A., 13
Einstein, A., 41-42, 48, 63, 71, 77
Emerson, R. W., 35, 74
Epicurus, 18-19

Index of Names

Euclid, 39

Faraday, M., 13, 88
Fichte, J. G., 79
Forman, P., 46
Foster, M. B., 70-71

Galileo, G., 27-28, 31, 38
Galvani, L., 13
Gardner, M., 30
Ghazzali, -al, 56
Gilson, E., 21, 73, 80
Giraud, V., 1
Glanvill, J., 22-23
Goblot, E., 37-38

Haldane, J. B. S., 82
Harris, R., 6
Hegel, G. W. F., 79
Heisenberg, W., 11, 45-47, 54
Hooke, R., 23-24
Horrocks, J., 23-24
Hume, D., 4, 17-21, 29, 68, 79
Huxley, T. H., 92
Huygens, C., 24

Jaki, S. L., 3, 10, 24-25, 28, 30, 32, 34, 37, 42-43, 47-49, 56, 71, 81, 86, 89-90
James, W., 79
Jefferson, T., 38-39
Jeremiah, 56
Jenkins, D., 66
Jung, C. G., 10

Kant, I., 4, 10-11, 43, 54
Kepler, J., 23-25, 27, 88
Kelsey, M., 5, 10
Knox, R., 7
Koyré, A., 25-26

Lamennais, F. R., 21
Landé, H., 77
Laplace, P. S., 32, 49, 83
Laurentin, R., 93
Leibniz, G. W., 24, 39-40
Lenin, V. L., 67
Lewis, C. S., 82-83

Mach, E., 79
MacKay, D. M., 58, 69-74
Marconi, G., 13
Margenau, H., 46
Maxwell, J. C., 10, 13, 41, 88
McCrea, W. H., 81
McGinley, L. J., 45
McInerny, R., 9
Mill, J. S., 28-29
Monden, L., 44
Montaigne, M. E., 3-4

Napoleon, 32
Newman, J. H., 7, 19
Newton, I., 24-29, 31, 33, 34

Ockham, W., 3, 29, 55-56, 70
Oersted, H. C., 13, 88
Ohm, G. S., 13
Olbers, W., 43
Olivieri, A., 93
Oresme, N., 90

Index of Names

Pascal, B., 4, 30
Paul VI, Pope, 9
Pauli, W., 48
Perrin, J., 51, 53
Pictet, M. A., 83
Pius X, Pope, Saint 2
Planck, M., 48-49
Plato, 28
Poincaré, H., 29, 37
Polanyi, M., 67-68
Pollard, W. G., 57-65, 67-68
Popper, K. R., 88

Rahner, K., 8
Ratzinger, J., 9, 11
Reid, T., 21
Resnick, R., 77
Robinson, J. A. T., 13, 15
Roentgen, W. C., 88
Rohrlich, F., 78
Rousseau, J. -J., 95

Sabourin, L., 87
Sabatier, A., 38
Sandage, A., 81
Schrödinger, E., 77
Séailles, G., 38-39
Seeberg, R., 95

Selvaggi, F., 44
Solovine, M., 43
Soubirous, Bernadette, Saint, 85, 93
Spinoza, B., 31-32, 79
Stokes, G. G., 12, 31, 34
Swinburne, R., 21

Taine, H., 1-2
Teilhard de Chardin, P., 85-86
Thomas, J. M. L., 7
Tolstoy, L., 66
Troyat, H., 66
Tschirnhausen, E. W. von, 32
Turner, J. E., 47, 60

Volta, A., 13
Voltaire, 21, 34-35, 62

Weber, W. E., 13
Werfel, F., 84-85
Wesley, J., 19
West, D. J., 87
Westfall, R. S., 23
Whitehead, A. N., 22

(continued from p. ii)

By the same author

Chance or Reality and Other Essays

*The Physicist as Artist:
The Landscapes of Pierre Duhem*

The Absolute beneath the Relative and Other Essays

The Savior of Science
(Wethersfield Institute Lectures, 1987)

God and the Cosmologists
(Farmington Institute Lectures, Oxford, 1988)

The Only Chaos and Other Essays

The Purpose of It All
(Farmington Institute Lectures, Oxford, 1989)

Catholic Essays

Cosmos in Transition: Studies in the History of Cosmology

Olbers Studies

Scientist and Catholic: Pierre Duhem

Reluctant Heroine: The Life and Work of Hélène Duhem

Universe and Creed

Genesis 1 through the Ages

Is There a Universe?

Patterns or Principles and Other Essays

Bible and Science

Theology of Priestly Celibacy

Means to Message: A Treatise on Truth

* * *

Translations with introduction and notes:
The Ash Wednesday Supper (Giordano Bruno)

Cosmological Letters on the Arrangement of the World Edifice (J.-H. Lambert)

Universal Natural History and Theory of the Heavens (I. Kant)

Note on the Author

Stanley L. Jaki, a Hungarian-born Catholic priest of the Benedictine Order, is Distinguished University Professor at Seton Hall University, South Orange, New Jersey. With doctorates in theology and physics, he has for the past thirty years specialized in the history and philosophy of science. The author of over thirty books and nearly a hundred articles, he served as Gifford Lecturer at the University of Edinburgh and as Fremantle Lecturer at Balliol College, Oxford. He has lectured at major universities in the United States, Europe, and Australia. He is honorary member of the Pontifical Academy of Sciences, *membre correspondant* of the Académie Nationale des Sciences, Belles-Lettres et Arts of Bordeaux, and the recipient of the Lecomte du Nouy Prize for 1970 and of the Templeton Prize for 1987.